GRANULAR MATERIALS

Edited by **Michael Sakellariou**

Granular Materials
http://dx.doi.org/10.5772/66007
Edited by Michael Sakellariou

Contributors

Francesco Federico, Kun Xue, Albert S. Kim, Abhinendra Singh, Sudeshna Roy, Hao Shi, Dallila Vescosci, Vanessa Magnanimo, Stefan Luding, Grasselli Yan, Bossis Georges, Meunier Alain, Volkova Olga, Emoke Imre, Michael Sakellariou

Published by InTech
Janeza Trdine 9, 51000 Rijeka, Croatia

Publishing Process Manager Mirena Calmic
Technical Editor SPi Global
Cover InTech Design team

First published September, 2017
Printed in Croatia
Legal deposit, Croatia: National and University Library in Zagreb

Additional hard copies can be obtained from orders@intechopen.com

Granular Materials, Edited by Michael Sakellariou
p. cm.
Print ISBN 978-953-51-3505-0
Online ISBN 978-953-51-3506-7

For EU product safety concerns:
IN TECH d.o.o., Prolaz Marije Krucifikse Kozulić 3, 51000 Rijeka, Croatia,
info@intechopen.com or visit our website at intechopen.com.

Meet the editor

Michael Sakellariou is a professor emeritus of Geomechanics and Engineering Structures at the National Technical University of Athens (NTUA). He studied Civil Engineering and Rural and Surveying Engineering at NTUA. He holds an MSc in Engineering Rock Mechanics from Imperial College London, and he obtained his PhD in Applied Mechanics from NTUA (1989). In his professional career, he was collaborator of engineering companies in major infrastructure projects. His teaching experience covers engineering mechanics, continuum mechanics, geotechnical engineering, soil mechanics and foundations, and engineering materials at undergraduate and postgraduate levels. His interests cover experimental mechanics, analytical and computational methods in geotechnical engineering, application of artificial intelligent and GIS in geotechnical engineering, structures monitoring using optical fiber sensors, and tectonic fault stress analysis.

Contents

Preface

Granular materials are a broad category of materials of interest in civil engineering, chemical engineering, physics, chemistry, and pharmacy. In engineering, great advances took place recently by the introduction, both, of experimental and computational methods. X-Ray tomography revealing the grain material structure and the distinct element computational method are two of them. The granular materials may behave as fluids or solids or both. The grain size may span from microscopic to macroscopic scale. From the wet sand effect, Reynolds inspired in 1885 the notion of granular universe introducing the term "dilatancy." In his Rede Lecture in Cambridge (1902) entitled *On an Inversion of Ideas as to the Structure of Universe,* he presented an explanation of the cause of electromagnetism giving a physical explanation of the equations linking electricity with magnetism. Later on, Lorentz gave the relativistic explanation of the electromagnetic waves. In 1930, Casagrande introduced the concept of critical void ratio to explain the liquefaction phenomenon. Donald Taylor in 1948 used the concept of dilatancy to explain the friction as a result of the interlocking of grains. Later on (1960s), Cambridge soil mechanics group (Roscoe, Schofield, Wroth, etc.) introduced the models of Cam-Clay and Granta-Gravel to explain the plastic behavior of soil. On the other hand, researchers from mathematics and physics like Bak, Tan, and Wisenfeld (1987, 1988) used the sand pile as a representative model of complex systems. Many researchers continue developing the concept of self-organized criticality to explain physical phenomena like earthquakes, landslides, and avalanches.

This volume presents basic notions and fundamental properties of granular materials covering a wide spectrum of granular material mechanics. In this collection of chapters, granular dynamics, granular flow from dilute to jammed states, dynamics of granular gas in microgravity, particle jetting induced by impulsive loadings, particle migration phenomena in embankment dams, and the grading entropy-based criteria of granular materials and filters are presented.

Dr. Michael Sakellariou
Professor Emeritus,
National Technical University of Athens,
Athens, Greece

Section 1

Introductory Chapter

Chapter 1

Introductory Chapter: A Short Survey of Landmarks

Michael G. Sakellariou

Additional information is available at the end of the chapter

http://dx.doi.org/10.5772/intechopen.70337

1. Introduction

Granular materials are a broad category of materials of interest in Civil engineering, Chemical engineering, Mining Industry, Physics, Chemistry, Pharmacy and Agriculture. In engineering, great advances took place recently by the introduction, both, of experimental and computational methods. X-ray tomography revealing the grain materials structure and the distinct elements computational method are two of them. The granular materials may behave as fluids or solids or both. The grains size may span from microscopic to macroscopic scale. From the wet sand effect, Reynolds inspired in 1885 [1] the notion of granular universe introducing the term "dilatancy." In his Rede Lecture in Cambridge [2] entitled *On an Inversion of Ideas as to the Structure of Universe*, he presented an explanation of the cause of electromagnetism giving a physical interpretation of the equations linking electricity with magnetism. In **Figure 1**, the experiment to show the effect of dilatation in the Rede Lecture is presented.

Later on, Lorentz gave the relativistic explanation of the electromagnetic waves. Schofield [3] gives a short account of Reynolds' ideas. In 1930, Casagrande introduced the concept of critical voids ratio to explain the liquefaction phenomenon. Donald Taylor in 1948 used the concept of dilatancy to explain the friction as result of the interlocking of grains. Later on in 1960, Cambridge soil mechanics group [4] introduced the models of Cam-Clay and Granta-Gravel to explain the plastic behavior of soil. On the other hand, researchers from Mathematics and Physics like Bak et al. [5] and Bak [6] used the sand pile as a representative model of complex systems (**Figure 2**).

Many researchers continue developing the concept of self-organized criticality to explain physical phenomena like earthquakes, landslides and avalanches. Werner [7], using notions and principles of Complexity Theory, explains the localization of shear: "...the most rapidly changing parts of the system tend to be localized in space. Familiar examples include localization of shear into narrow bands in turbulent fluids and in solid deformation or fracture. Patterns inherently involve localization because they can be described with fewer spatial dimensions than the space in which they are embedded: lines for two-dimensional patterns and surfaces in three-dimensional patterns."

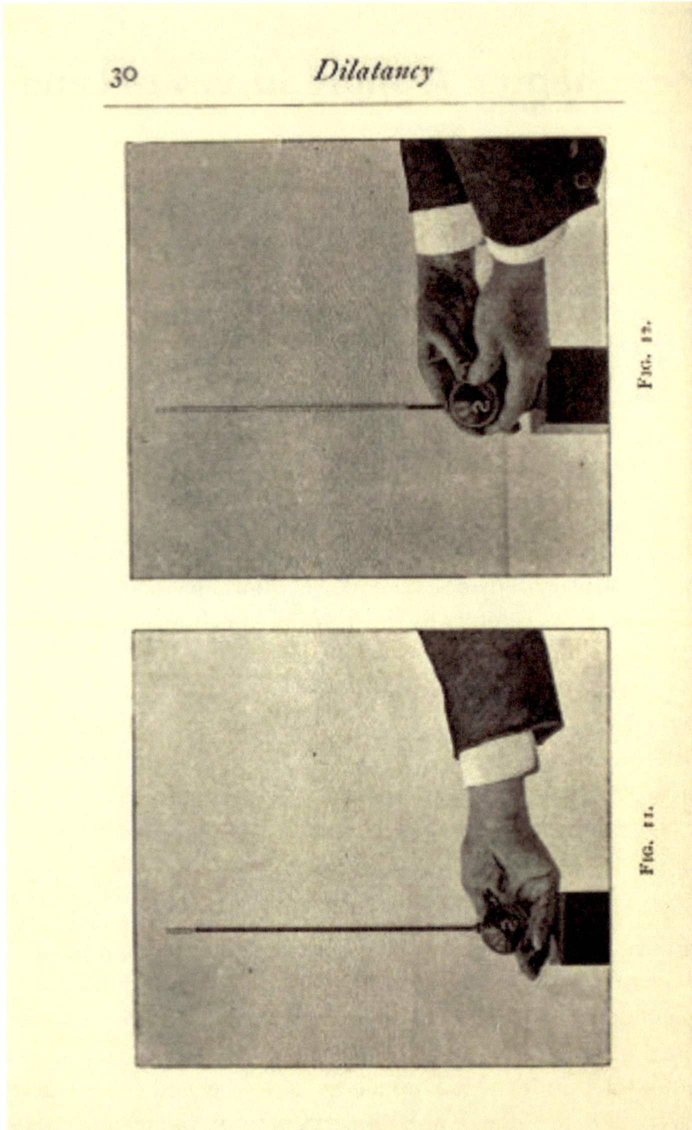

Figure 1. Experiment presented in Reynolds' Rede Lecture to show the dilatancy phenomenon. In Fig. 11 of the original publication, a bag filled with water is shown. After squeezing the bag, the water rises in the tube. In contrary, in Fig. 12, as above, a big filled with small shot and colored water is shown. In this case, the squeezing of the bag caused a drop of the water in the tube resulted from the increase of voids in the coarse material.

This volume presents basic notions and fundamental properties of granular materials covering a wide spectrum of granular material mechanics. In this collection of chapters, granular dynamics, granular flow from dilute to jammed states, dynamics of granular gas in micro-gravity, particle jetting induced by impulsive loadings, particle migration phenomena in embankment dams and the grading entropy-based criteria of granular materials and filters are presented. The scientific areas of the above are very broad, with applications from earth dams to space flights.

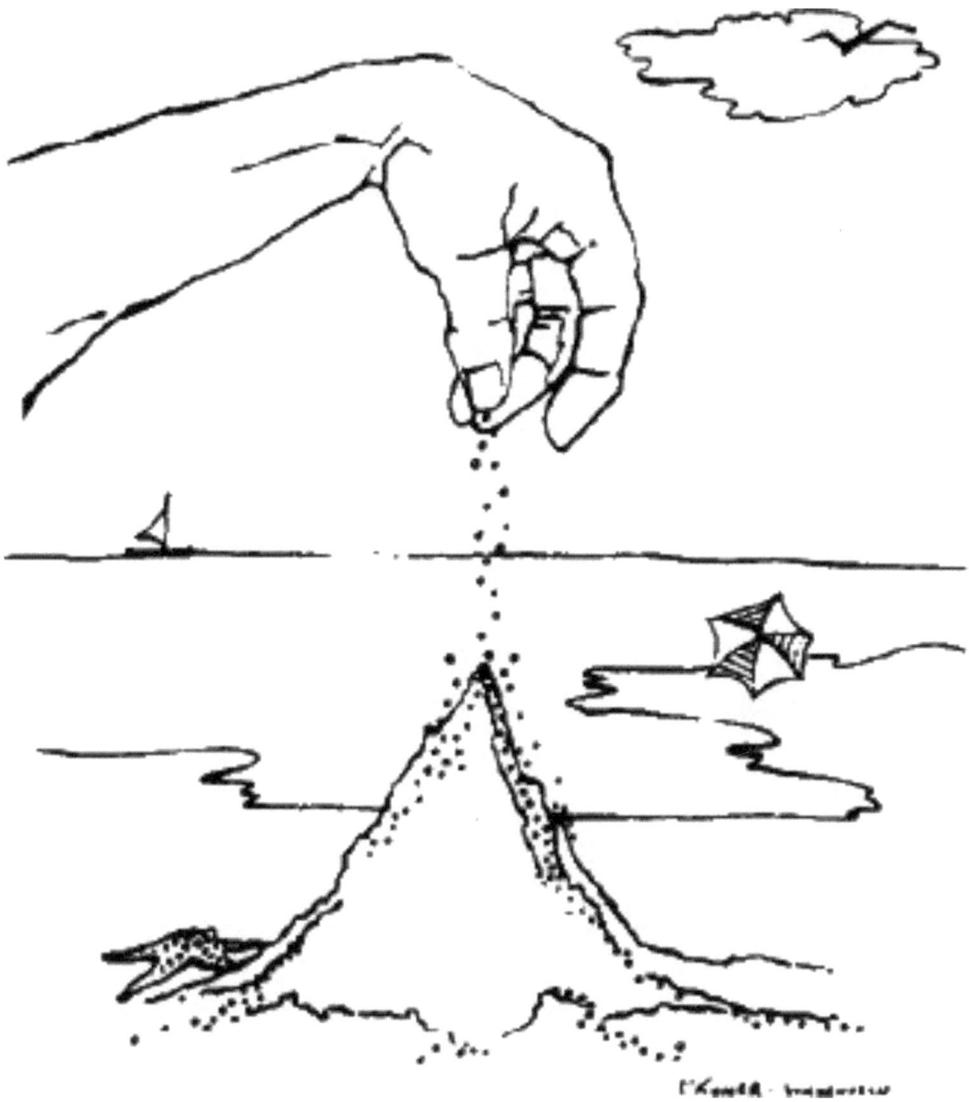

Figure 2. The sand pile as the prototypical example of the self-organized criticality [6].

In the first chapter, a theoretical framework of granular dynamics using a generalized Lagrangian approach is presented, discussing future directions of computational granular dynamics. The second chapter focuses on the mechanical responses of a granular material describing the mechanical behavior of dissipative, deformable particles in different states, from fluid to solid states. The next chapter presents an experimental study of the dynamical behavior of a model granular medium. The model has been submitted to external vibrations under microgravity, the whole process being recorded using high-speed video camera. The fourth chapter is an experimental and numerical study of the particle jetting phenomenon observed in the explosive or shock disposal of particle rings, shells or cylinders. Continuum approach, for the explosive case, and Distinct Element Method for the shock-induced jetting have been adopted. In the following two chapters, important hydraulic properties of earth dams are presented, affecting their structural integrity. First, a numerical procedure has been developed to simulate particle migration phenomena due to seepage resulting erosion. The method has been applied in case studies. The final chapter presents entropy-based criteria to control the internal erosion process in earth dams. According to this procedure, the whole grading curve is used instead of a limited number of points, as is usually adopted.

Author details

Michael G. Sakellariou

Address all correspondence to: mgsakel@mail.ntua.gr

National Technical University of Athens, Greece

References

[1] Reynolds O. On the dilatancy of media composed of rigid particles in contact, with experimental illustrations. Philosophical Magazine Series 5. 1885;20(127):469-481

[2] Reynolds O. On an Inversion of Ideas as to the Structure of the Universe. The Rede Lecture 1902. Cambridge University Press, Cambridge; 1902

[3] Schofield A. Disturbed Soil Properties and Geotechnical Design. Thomas Telford Ltd, London; 2005. p. 33

[4] Roscoe KH. Tenth Rankine Lecture: The Influence of Strains in Soil Mechanics. Géotechnique. 1970;20(2):129-170

[5] Bak P, Tang C, Wiesenfeld K. Self-organized criticality: An explanation of 1/f noise. Physical Review Letters. 1987;59(4):381-384

[6] Bak P. Self-organized criticality. Physica A. 1990;163:403-409

[7] Werner BT. Complexity in natural landform patterns. Science. 1999;284(5411):102-107

Section 2

Fundamental Concepts of Granular Materials Behaviour

Chapter 2

Dissipative Dynamics of Granular Materials

Albert S. Kim and Hyeon-Ju Kim

Additional information is available at the end of the chapter

http://dx.doi.org/10.5772/intechopen.69196

Abstract

Granules are inelastic particles, undergoing dissipative and repulsive forces on contact. A granular state consists of a conglomeration of discrete, non-Brownian particles in a combined state of solid, liquid, and gas. Modern theoretical physics lacks general theories for the granular states. Simulation methods for particle dynamics include molecular dynamics (MD), Brownian dynamics (BD), Stokesian dynamics (SD), dissipative particle dynamics (DPD), and dissipative hydrodynamics (DHD). These conventional methods were originally designed to mimic the small-particle motion being less influenced by the gravitational force. There are three reasons that a conventional method cannot be directly applied to investigate granular dynamics. First, volume exclusion forces between colliding particles are often disregarded due to strong repulsive forces between negatively charged colloids and nanoparticles. Second, the gravitational force is not significant as applied to small, light particles, and therefore it is often discarded in force/torque calculations. Third, energy conservation in an equilibrium state is not guaranteed for the granular system due to the inelastic and frictional nature of the granular materials. In this light, this chapter discusses the fundamentals of particle dynamics methods, formulates a robust theoretical framework for granular dynamics, and discusses the current applications and future directions of computational granular dynamics.

Keywords: granular dynamics, least action principle, classical mechanics, Newton's law of motion, Hertz's law, inelasticity, compression, restitution coefficient, parallel particle dynamics, dissipative hydrodynamics

1. Introduction

1.1. Mechanics

Mechanics is the investigation of physical bodies when they are subjected to forces and torques in Euclidean three-dimensional spaces. It is often referred to as classical mechanics (in physics), which is closely related to engineering mechanics. Classical mechanics has two branches: statics and dynamics. *Statics* is concerned with the equilibrium of a body that is either at rest or moves with a constant velocity. *Dynamics* deals with the accelerated motion of a body, classified into two parts: *kinematics*, which treats only the geometric aspect of motion, and *kinetics*, which analyzes the forces causing the motion. Two representative objects in mechanics are a particle and a rigid body. A particle is the most basic unit of matter, which contains a mass of a negligible volume. Since the particle is small enough to be regarded as a point mass, its angular motion is completely discarded in analyzing its dynamics. The total energy of particles depends on their velocities and positions as influenced by external and internal forces. The mass of each particle is assumed invariant and therefore energy conservation is independent of mass conservation in classical mechanics. If particles of interest have sub-atomic sizes (such as hydrogens and electrons), classical mechanics fails to predict their intrinsic duality behaviors. Quantum mechanics explains matter's simultaneous wave-like and particle-like properties, and unifies matter and energy as they converge at the level of Planck's constant. A particle's position and velocity cannot be measured accurately at a particular moment because if one measures the position accurately, then the particle's momentum will be disturbed, and vice versa. This is called the Heidelberg uncertainty principle. In the macroscopic engineering world, in which humans observe objects with their naked eyes, quantum phenomena are extremely rare to observe. Although quantum mechanics includes classical mechanics as a sub-set, most conventional engineering phenomena are macroscopic enough to neglect sub-atomic effects. Based on the above-mentioned characteristics and classifications of mechanics, the granular dynamics in this chapter focuses on kinetics and hydrodynamics of multiple non-Brownian particles in locally confined three-dimensional (3D) spaces, often filled with fluid media.

1.2. Principles in classical mechanics

1.2.1. Governing equations

Classical mechanics in this chapter is narrowly defined as the investigation of the motion of systems of bodies under the influence of forces and torques. The problem is to determine the positions of all the particles at an arbitrary time t from their initial state at $t = 0$. Newton's laws for the motion of bodies can be summarized as follows:

1. Newton's first law states that an object will remain at rest or in linear motion unless acted upon by an external force. (This first law describes a constant velocity motion in the absence of an external force, which is a special case of zero acceleration in the second law below.)

2. Newton's second law states that the net (unbalanced) force $\boldsymbol{F}$ on an object is equal to the product of the mass m and acceleration $\boldsymbol{a}$ of the object:

$$\boldsymbol{F} = m\boldsymbol{a} \tag{1}$$

where the acceleration is defined as the rate of change in velocity with respect to time. (This law indicates that the net force modifies the object's velocity with respect to time. Then, the mass m can be interpreted as the object's resistance to the velocity change.)

3. Newton's third law states that all the forces in the universe occur in equal (in magnitude) but opposite directions. For example, when one body exerts a force on a second body, the second body simultaneously exerts a force, on the first body, equal in magnitude and opposite in direction. These two forces cancel each other, and therefore, the net sum is always zero, even if particles are inelastic.

As noted above, particle size is neglected in describing its motion. The possibility of doing so depends on the actual size of the object and/or its distance from the observer. But, when a group of constrained particles forms a rigid body, its rotational motion is described using an equation similar to Newton's second law of force, which is

$$\boldsymbol{M} = I\boldsymbol{\alpha} \tag{2}$$

where $\boldsymbol{M}$ is the torque or moment, I is the mass moment of inertia, and α is the angular acceleration. A rigid body has six degrees of freedom: three for translation and the other three for rotation. One can combine Eqs. (1) and (2) to write the governing equation of motion of a rigid body in a simple matrix form:

$$\begin{bmatrix} \boldsymbol{F} \\ \boldsymbol{M} \end{bmatrix} = \begin{bmatrix} M & 0 \\ 0 & I \end{bmatrix} \begin{bmatrix} \boldsymbol{a} \\ \boldsymbol{\alpha} \end{bmatrix} \tag{3}$$

where the zeros in the off-diagonal terms indicate that the medium in which particles are moving is not viscous, i.e., conceptually similar to vacuum. The linear and angular accelerations are defined as

$$\boldsymbol{a} = \ddot{\boldsymbol{r}} = \frac{d^2\boldsymbol{r}}{dt^2} \quad \text{and} \quad \boldsymbol{\alpha} = \ddot{\boldsymbol{\theta}} = \frac{d^2\boldsymbol{\theta}}{dt^2} \tag{4}$$

respectively, where $\boldsymbol{r}$ and $\boldsymbol{\theta}$ are the linear and angular positions of the object, of which time derivatives are the linear and angular velocities, respectively:

$$\boldsymbol{v} = \dot{\boldsymbol{r}} = \frac{d\boldsymbol{r}}{dt} \quad \text{and} \quad \boldsymbol{\omega} = \dot{\boldsymbol{\theta}} = \frac{d\boldsymbol{\theta}}{dt} \tag{5}$$

Where $\boldsymbol{a}$ and $\boldsymbol{\alpha}$ have three components each so that $[\boldsymbol{a}, \boldsymbol{\alpha}]^T$ in Eq. (3) is a vector of six components, where the superscript T indicates a transpose. For mathematical simplicity, a generalized coordinate $\boldsymbol{q}$, generalized velocity $\dot{\boldsymbol{q}}$, and generalized force $\boldsymbol{Q}$ of an object are written as

$$\boldsymbol{q} = \begin{bmatrix} \boldsymbol{r} \\ \boldsymbol{\theta} \end{bmatrix}, \quad \dot{\boldsymbol{q}} = \begin{bmatrix} \dot{\boldsymbol{r}} \\ \dot{\boldsymbol{\theta}} \end{bmatrix}, \quad \text{and} \quad \boldsymbol{Q} = \begin{bmatrix} \boldsymbol{F} \\ \boldsymbol{M} \end{bmatrix} \tag{6}$$

The classical mechanics problems are usually reduced to solve for $\boldsymbol{q}(t)$ and $\dot{\boldsymbol{q}}(t)$ at time t under the influence of specified $\boldsymbol{Q}$, using the initial conditions of $\boldsymbol{q}(t=0)$ and $\dot{\boldsymbol{q}}(t=0)$.

1.2.2. *Total energy sum and difference*

Fundamental questions from physicists are "What governs the motion of an object?" and "How can the motion be described and predicted mathematically?" Then, a fundamental question that naturally rises is "Is there a more fundamental principle that nature follows other than Newton's second law?"

Let's consider a particle, i.e., a point mass, found at position $\boldsymbol{r}(t)$ with velocity $\boldsymbol{v}(t)$ under the influence of force $\boldsymbol{f}(\boldsymbol{r})$, depending on the particle position only. We consider Newton's second law in one-dimensional space:

$$f(x) = m\frac{dv}{dt} \tag{7}$$

Because $dx = vdt$, we multiply dx with $f(x)$ and vdf with $mdv\,/\,dt$ to derive

$$f(x)\,dx = mv\;dv \tag{8}$$

and integrate each side from state 1 of (x_1, v_1) to state 2 of (x_2, v_2) to obtain

$$\int_{x_1}^{x_2} f(x)\;dx = m\int_{v_1}^{v_2} vdv = \frac{1}{2}mv_2^2 - \frac{1}{2}mv_1^2 \tag{9}$$

Then, Eq. (9) can be rewritten as

$$\Delta W = T_2 - T_1 \tag{10}$$

where $\Delta W = \int_{x_1}^{x_2} f(x)dx$ is a *work done* and $T = \frac{1}{2}mv^2$ is the *kinetic energy*. Eq (10) indicates that the work done is equal to the kinetic energy change from states 1 to 2. The integration of force with respect to x in Eq. (9) can be exact if the force depends on the particle position only. If so, this force is called *conservative* and becomes the satisfactory condition for the energy conservation principle. A conservative $f(x)$ can be expressed as a gradient of a scalar function V:

$$f(x) = -\frac{dV(x)}{dx} \quad \text{in 1-D} \tag{11}$$

$$\boldsymbol{f} = -\nabla V(\boldsymbol{r}) \quad \text{in 3-D} \tag{12}$$

where V is called the *potential energy function*. Then, the force integration is simply

$$\int_{x_1}^{x_2} f(x)\,dx = -\int_{x_1}^{x_2} dV = -V(x_2) + V(x_1) \tag{13}$$

Substitution of Eq. (13) into Eq. (9) gives

$$T_1 + V_1 = T_2 + V_2 \tag{14}$$

$$E_1 = E_2 = \text{constant} \tag{15}$$

where total energy E defined as a sum of the potential and kinetic energies, i.e., $E = T + V$, is derived as constant regardless of the path the particle takes. The potential energy difference depends on only the initial and final positions of x_1 and x_2, respectively. Note that the total energy is conserved only if the force depends on only particle location. In advanced classical mechanics, the total energy E is replaced by a Hamiltonian function: $H = T + V$, and problems can be solved identically to applying Newton's second law. Using the Legendre transformation of the Hamiltonian H, a new function called Lagrangian L is defined as the difference between the kinetic and potential energies:

$$L = T - V \tag{16}$$

Instead of dealing with force vectors in Newton's second law, Lagrangian mechanics uses the scalar Lagrangian function, which is assumed to contain all the information of the mechanical system.

1.2.3. *Principle of the least action*

The most general formulation of the law governing the motion of mechanical systems is *the principle of least action* or *Hamilton's principle* [1]. According to this principle, a mechanical system is characterized using a definite Lagrangian function $L(q_1, q_2, \ldots, q_s, \dot{q}_1, \dot{q}_2, \ldots, \dot{q}_s, t)$ or briefly $L(q, \dot{q}, t)$, and the motion of the system is such that a certain condition (discussed below) is satisfied.

At time t_1 and t_2, particle positions are defined by two sets of the generalized coordinates, $q(t_1)$ and $q(t_2)$. The condition is that the system moves between these two positions, minimizing the integral

$$S = \int_{t_1}^{t_2} L(q, \dot{q}, t)\mathrm{d}t \tag{17}$$

to the least possible value. The integral of Eq. (17) is called the *action*. Note that the Lagrangian contains generalized coordinates and velocities, q and $\dot{q}$ only (not the higher derivatives such as $\ddot{q}$), as independent variables.

Let us now derive the differential equations that minimize the action integral of Eq. (17). For simplicity, the system is assumed to have only one degree of freedom. Let $q = q(t)$ be the function for which the action S is a minimum. This means that S changes when $q(t)$ is replaced by

$$q(t) + \delta q(t) \tag{18}$$

where $\delta q(t)$, called a variation of $q(t)$, is a function, assumed to be small everywhere in the interval of time from t_1 to t_2. Since Eq. (18) must include the values of $q(t_1)$ and $q(t_2)$, we can now conclude that

$$\delta q(t_1) = \delta q(t_2) = 0 \tag{19}$$

In this case, the change in S when q is replaced by $q + \delta q$ is equal to

$$\int_{t_1}^{t_2} L(q + \delta q, \dot{q} + \delta\dot{q}, t) - \int_{t_1}^{t_2} L(q, \dot{q}, t) \tag{20}$$

We expand the difference in powers of δq and $\delta\dot{q}$ in the integrand and leave only the first-order terms. Then, the principle of least action may be written in the form

$$\delta S = \delta\int_{t_1}^{t_2} L(q, \dot{q}, t) = 0 \tag{21}$$

or equivalently

$$\int_{t_1}^{t_2} \left(\frac{\partial L}{\partial q}\delta q + \frac{\partial L}{\partial \dot{q}}\delta\dot{q}\right)\mathrm{d}t = 0 \tag{22}$$

Since $\delta\dot{q} = \mathrm{d}\delta q/\mathrm{d}t$, we integrate the second term of Eq. (22) by parts to obtain

$$\delta S = \left[\frac{\partial L}{\partial \dot{q}}\delta q\right]_{t_1}^{t_2} + \int_{t_1}^{t_2}\left(\frac{\partial L}{\partial q} - \frac{\mathrm{d}}{\mathrm{d}t}\frac{\partial L}{\partial \dot{q}}\right)\delta q\mathrm{d}t \tag{23}$$

The condition of Eq. (19) shows that the integrated term in Eq. (23) is zero:

$$\left[\frac{\partial L}{\partial \dot{q}}\delta q\right]_{t_1}^{t_2} = 0 \tag{24}$$

and the remaining integral is

$$\int_{t_1}^{t_2}\left(\frac{\partial L}{\partial q} - \frac{\mathrm{d}}{\mathrm{d}t}\frac{\partial L}{\partial \dot{q}}\right)\delta q\mathrm{d}t = 0, \tag{25}$$

which must vanish for all values of δq. This can be satisfied if and only if the integrand of Eq. (25) is zero. Thus, we have

$$\frac{\mathrm{d}}{\mathrm{d}t}\left(\frac{\partial L}{\partial \dot{q}}\right) - \frac{\partial L}{\partial q} = 0 \tag{26}$$

When the system has more than one degree of freedom, then Eq. (26) becomes

$$\frac{\mathrm{d}}{\mathrm{d}t}\left(\frac{\partial L}{\partial \dot{q}_i}\right) - \frac{\partial L}{\partial q_i} = 0 \quad (i = 1, 2, \ldots, s) \tag{27}$$

where s is the total degrees of freedom of the particles in the system. Eq. (27) is a set of required differential equations, called in mechanics *Lagrange's equations*. If there is no constraint, the total degrees of freedom of a system containing N_p objects, s, is equal to $6N_p$.

The zero on the right-hand side of Eq. (27) implies that the total energy is conserved because particle forces depend on their positions only. There are no forces/torques that dissipate the energy. If a number of particles are investigated under the influence of dissipative forces, such as hydrodynamic drag and frictional forces, Eq. (27) should be modified to

$$\frac{\mathrm{d}}{\mathrm{d}t}\left(\frac{\partial L}{\partial \dot{q}_i}\right) - \frac{\partial L}{\partial q_i} = Q_i^{\dagger} \quad (i = 1, 2, \ldots, s). \tag{28}$$

where $Q^{\dagger}$ is the generalized non-conservative force. Therefore, we take Eq. (28) as the general governing equation of motion for multi-body granular dynamics in the regime of classical mechanics and microhydrodynamics [2]. It generalizes Newton's second law and usually provides great mathematical simplicity by dealing with a scalar function L instead of force vectors.

2. Particle dynamics simulation methods

In science and engineering disciplines, dynamic simulations of particulate materials allow researchers to investigate microscopic many-body phenomena and further predict macroscopic material properties. In a liquid (aqueous) phase, rigorous and accurate simulations of particle dynamics must consider the repulsive volume-exclusion between polydispersed particles. Here, we briefly review the historical development of particle dynamics simulation methods in various scales.

2.1. Molecular dynamics

Conventional molecular dynamics (MD) treats particles (such as ions and molecules) as interacting point-masses and updates their present positions and velocities (at time t) to those in the future (at time $t + dt$) [3]. Basic potential forms include the hard-sphere and Lennard-Jones potential [4]. A specific analysis of particle trajectories can provide measurable macroscopic quantities such as solute diffusivities in an aqueous medium and heat capacities in various thermal conditions. The choice of time internal dt must be much smaller than the typical time for a molecule to travel a distance of the same order as its size. One of the most widely used methods to integrate the governing Eq. (1) is the Verlet algorithm [5], which provides a direct solution of the second-order ordinary differential equation with errors of δt^4 order. The particle position at time $t + \delta t$ can be expanded at time t using Taylor's series:

$$\boldsymbol{r}(t + \delta t) = \boldsymbol{r}(t) + \boldsymbol{v}(t)\delta t + \frac{1}{2}\boldsymbol{a}(t)\delta t^2 + O(\delta t^3) \tag{29}$$

and by replacing $+\delta t$ with $+\delta t$, one obtains

$$\boldsymbol{r}(t - \delta t) = \boldsymbol{r}(t) - \boldsymbol{v}(t)\delta t + \frac{1}{2}\boldsymbol{a}(t)\delta t^2 - O(\delta t^3) \tag{30}$$

The position $\boldsymbol{r}$ and velocity $\boldsymbol{v}$ at the future time $t + \delta t$ can be calculated by adding and subtracting Eqs. (29) and (30) such as

$$\boldsymbol{r}(t+\delta t)=2\boldsymbol{r}(t)-\boldsymbol{r}(t-\delta t)+\delta t^2\boldsymbol{a}(t) \tag{31}$$

and

$$v(t)=\frac{\boldsymbol{r}(t+\delta t)-\boldsymbol{r}(t-\delta t)}{2\delta t} \tag{32}$$

respectively. As the primary objective of MD is to evolve the particle system, one of the advantages of the Verlet algorithm of Eq. (31) is that the velocity does not need to be calculated unless needed during simulations. This advantage does not exist if hydrodynamic drag forces are considered, the magnitude of which increases with respect to the relative velocity of particles to that of the fluid medium. The particle position at the next time step is calculated using those of the past and current time steps and the acceleration vector $\boldsymbol{a}$ obtained by the net force exerted on the particles. Velocity vectors can be additionally calculated during or after simulations if the kinetic energy needs to be calculated. Advanced integration algorithms over Verlet's algorithm include the half-step leap-frog scheme [6] and velocity-Verlet algorithm [7]. Several leading MD simulation packages include assisted model building and energy refinement (AMBER) [8], chemistry at HARvard macromolecular mechanics (CHARMm) [9], Groningen machine for chemical simulations (GROMACS) [10], and nanoscale molecular dynamics (NAMD) [11]. The main difference between these MD simulation packages is how the force fields (functional forms and parameter values) are determined and used as parameters in potential functions. Applications of MD simulations are diverse and can be used in a wide variety of chemical and environmental applications. The pre-developed time evolution algorithms can be used in most simulations of particle dynamics, governed by Newton's second law. On the other hand, a direct application of MD for granular dynamics simulation is limited to mimicking granular phenomena and properties because MD was originally developed for particles treated as point masses.

2.2. Brownian dynamics

Experimental and theoretical studies on Brownian motion were initiated by Brown [12], Einstein [13], Langevin [14], and Chandrasekhar [15, 16]. When solute motion in a solution is of greater interest, the motion of the tremendous number of ambient solvent molecules cannot be or does not need to be computed. Therefore, effects of solvent motion are replaced by random, fluctuating forces acting on solutes. Then, the governing equation is switched from Newton's Eq. (1) to Langevin's equation:

$$m\frac{d\boldsymbol{v}}{dt}=-\xi\boldsymbol{v}+\boldsymbol{f}(\boldsymbol{r})+\boldsymbol{f}'(t) \tag{33}$$

where $-\xi v$ is the hydrodynamic drag force acting in the opposite direction to the relative velocity of particle in the fluid medium, ξ is the drag coefficient (to be expanded to a tensor form), $\boldsymbol{f}(\boldsymbol{r})$ is the conservative force derived from the potential energy function, and $\boldsymbol{f}'(t)$ is the random fluctuation force caused by adjacent, rapidly-moving solvent molecules. It is assumed that $\boldsymbol{f}'$ is a Gaussian process with infinitely small correlation time, which renders the famous fluctuation-dissipation theorem [17]

$$\langle f'(t_1) \cdot f'(t_2) \rangle = 2\xi k_B T \delta(t_1 - t_2) \tag{34}$$

$$\langle f'(t) \rangle = 0 \tag{35}$$

After a sufficiently long time after the initial state which ensures Eq. (35) is true, one can approximate Eq. (33) in the absence of f as

$$\left\langle \frac{dv}{dt} \right\rangle \simeq -\left\langle \frac{v}{m/\xi} \right\rangle \simeq -\frac{\langle v \rangle}{\tau} \tag{36}$$

where $\langle \cdots \rangle$ indicates averages over time. Eq. (36) allows us to determine the time scale of the decelerating motion, the so-called relaxation time, defined as $\tau = m / \xi$. The time step dt in Brownian dynamics (BD) should be much longer than the particle relaxation time, i.e., $dt \gg \tau$. Only for a dilute solution, the drag coefficient ξ can be regarded as a constant. Stokes derived the drag coefficient of a spherical particle moving in a fluid medium as $\xi = 3\,\pi d_p \mu$, where d_p is the (hydrodynamic) radius of the particle and μ is the absolute viscosity of the fluid medium [18]. In general, Stokes-Einstein diffusivity is defined as

$$D_{\text{SE}} = \frac{k_B T}{\xi} = \frac{k_B T}{3\pi \mu d_p} \tag{37}$$

Although BD adopts the Oseen tensor, it still treats a particle as a point mass (like MD). From BD simulations, the hydrodynamic diameter d_p can be inversely determined by matching experimental data and simulation results. As a consequence, the volume-exclusion based on particle sizes is commonly disregarded in MD as well as BD simulations [19–23]. Specific features of MD and BD as applied to colloidal systems can be found elsewhere [24].

2.3. Dissipative particle dynamics

The restrictive condition of the time interval ($dt \gg \tau$) is fundamentally resolved in dissipative particle dynamics (DPD) for finite-sized particles by incorporating the Fokker-Planck equation and the Ito-Wiener process [25–31]. The total force on particle i from other particles is written in the form of:

$$m\boldsymbol{a}_i = \sum_{j=1, j\neq i}^{N_p} \left(\boldsymbol{F}_{ij}^{P} + \boldsymbol{F}_{ij}^{D} + \boldsymbol{F}_{ij}^{R} \right) \tag{38}$$

where $\boldsymbol{F}^P$ is a conservative inter-particle force, $\boldsymbol{F}^D$ is a dissipative force, and $\boldsymbol{F}^R$ is a random fluctuation force. It is required that $\boldsymbol{F}^D$ and $\boldsymbol{F}^R$ are linear and independent of the momentum, respectively. A simple form of these force are hypothesized as

$$\boldsymbol{F}_{ij}^{D} = -\gamma \omega_D(r_{ij})(\boldsymbol{e}_{ij} \cdot \boldsymbol{v}_{ij})\boldsymbol{v}_{ij} \tag{39}$$

$$\boldsymbol{F}_{ij}^{R} = \sigma \omega_R(r_{ij})\boldsymbol{e}_{ij}\zeta_{ij} \tag{40}$$

where $r_{ij} = |\boldsymbol{r}_i - \boldsymbol{r}_j|$, $\boldsymbol{v}_{ij} = \boldsymbol{v}_i - \boldsymbol{v}_j$, $\boldsymbol{e}_{ij} = (\boldsymbol{r}_i - \boldsymbol{r}_j)/r_{ij}$, and $\sigma = \sqrt{2k_B T \gamma}$. Importantly, $\zeta_{ij} = \zeta_{ji}$ is a Gaussian white-noise ther of zero mean and unit variance. The stochastic differential equation (SDE) of DPD consists of

$$dr_i = v_i dt \tag{41}$$

$$m dv_i = \sum_j [F_{ij}^P + F_{ij}^D] dt + \sum_j \sigma \omega_R(r_{ij}) e_{ij} dW_{ij} \tag{42}$$

where dW_{ij} (= dW_{ji}) is independent increment and of the Wiener process, satisfying

$$dW_{ij} dW_{kl} = (\delta_{ik}\delta_{jl} + \delta_{jl}\delta_{jk}) dt \tag{43}$$

i.e., $dW_{ij}(t)$ is an infinitesimal of order 1/2 (i.e., proportional to $\sqrt{\delta t}$).

In Eq. (42), the momentum changes due to the conservative and dissipative forces are proportional to dt, and due to the random fluctuating, force is linear on $\sqrt{dt}$. Hydrodynamic resistances are presumed to be pairwise and described as (intuitively chosen) simple functions of the inter-particle distance r:

$$\omega_D = \omega_R^2 = \begin{cases} (1 - r/r_c)^2 & \text{for} \quad r < r_c \\ 0 & \text{otherwise} \end{cases} \tag{44}$$

where r_c is the cut-off distance. For two particles in close proximity, it was reported that the pairwise summation of hydrodynamic forces can be erroneous in estimating the many-body hydrodynamic forces/torques, represented using fourth-order tensors [32]. Interestingly, DPD was widely used to investigate the motion of macromolecules and polymers, which are soft and deformable, in liquid phases [25–31, 33–35]. The computational advantages of DPD include that, first, the tensor-wise hydrodynamic interactions are simplified to pairwise forces (although this approach has less fundamental rigor); second, the smooth functional form of Eq. (44) allows multiple soft particles to physically overlap each other, in which repulsive lubrication forces are disregarded; and third, the fluctuation-dissipation theorem is automatically satisfied [17] so that the thermodynamics of the system are well described.

2.4. Stokesian dynamics

The lubrication tensors in Stokesian dynamics (SD) may mimic the volume exclusion forces, which is logarithmically proportional to the surface-to-surface distance between two particles [36]. When two spheres are colliding or in contact with each other, the surface-to-surface distance converges to zero and the lubrication force diverges to infinity. In SD, a many-body, far-field, grand mobility matrix $\mathcal{M}^\infty$ is built using the pairwise superposition of the two-body mobility matrix. Its product with the hydrodynamic force F^H gives the relative velocities of particles to the fluid flow, ΔU [37–41]:

$$\Delta U = [\mathcal{M}^\infty] \cdot F^H \tag{45}$$

where the superscript ∞ indicates that the near-field lubrication forces are excluded. The grand resistance matrix is formed by inverting the grand mobility matrix and correcting the near-field lubrication such as

$$[\mathcal{R}] = [\mathcal{M}^{\infty}]^{-1} - [\Delta\mathcal{R}_{2B}] \quad (46)$$

where $[\Delta\mathcal{R}_{2B}]$ indicates a pairwise addition of the exact two-body resistance matrix $\mathcal{R}_{i,j}^{(2)}$ subtracted by the inverse of two-body far-field mobility matrix, $\mathcal{M}_{ij}^{(2)}$:

$$[\Delta\mathcal{R}_{2B}] = \sum_{i,j} \left(\mathcal{R}_{i,j}^{(2)} - [\mathcal{M}_{ij}^{(2)}]^{-1} \right) \quad (47)$$

SD provides the many-body diffusion tensor, which has a significantly higher accuracy than those of BD and DPD. SD is, however, fundamentally limited to rigid particles of no contact. Similar to BD, SD uses the Langevin equation of the drag force (Eq. (33)) and therefore is subjected to the intrinsic restriction of the time interval, $dt \gg \tau$. Most of the SD simulations often deal with zero-inertia motion so that only *drifting* motion of particles is investigated with no-acceleration under conservative and external force fields in a viscous fluid media. To generally mimic the complex hydrodynamic motion of many particles in a fluid medium, the zero-force assumption should be relaxed and the time step dt is arbitrarily chosen for computational efficiency under given physical conditions. Recent studies on SD include fast numerical inversion of the grand mobility matrix to accelerate the computational time, but the hydrodynamics is similarly mimicked [42, 43].

2.5. Dissipative hydrodynamics

Dissipative hydrodynamics (DHD) was recently developed to unify the above-mentioned particle dynamics methods [44]. DHD employed specific advantageous features from various simulation methods, specifically the many-body hydrodynamic tensors from SD and the stochastic differential equation from DPD. DHD can mimic the translational as well as rotational motion of N_p particles in a viscous fluid of temperature T. Importantly, it is free from the restriction of the particle relaxation time. The stochastic governing equations of DHD are represented as

$$\boldsymbol{M} \cdot d\boldsymbol{u} = [\boldsymbol{Q}^p - \boldsymbol{R} \cdot (\boldsymbol{u} - \boldsymbol{U})]dt + \boldsymbol{B} \cdot d\boldsymbol{W} = \boldsymbol{Q}\,dt \quad (48)$$

where $\boldsymbol{M}$ is the mass/moment-of-inertia matrix of $6N_p \times 6N_p$ dimension, $\boldsymbol{u}$ and $\boldsymbol{U}$ are translational/rotational velocities of particles and the fluid, respectively, $\boldsymbol{Q}^p$ is the conservative force/torque vector, $\boldsymbol{R}$ is the grand resistance matrix, $\boldsymbol{B}$ is the Brownian matrix of zero mean and finite variance, i.e., $\langle \boldsymbol{B} \rangle = 0$ and $\langle \boldsymbol{B}^{\mathrm{tr}} \cdot \boldsymbol{B} \rangle = 2k_BT$, where k_B is the Boltzmann constant, and $d\boldsymbol{W}$ is the Ito-Wiener process of $6N_p$ elements [45, 46]. The generalized force/torque is defined, similar to Eq. (42), as

$$\boldsymbol{Q} = \boldsymbol{Q}^p - \boldsymbol{R} \cdot (\boldsymbol{v} - \boldsymbol{U}) + \boldsymbol{B} \cdot \boldsymbol{w}dt \quad (49)$$

where $\boldsymbol{w} = d\boldsymbol{W}/\sqrt{dt}$. A simple time integration using an intermediate time step is adopted from standard DPD algorithms [31]. It is worth noting that the SDE (Eq. (48)) relaxes the intrinsic time interval restriction of $dt \gg \tau$, which BD and SD are subjected to. Therefore, DHD allows mimicking accelerated motion of particles of different sizes. Mathematical details of the DHD formalism can be found elsewhere [44]. Applications of DHD and its related work in environmental engineering include, but are not limited to, aggregate formation [47, 48] and single collector granular filtration [49]. DHD provides the most fundamental and accurate simulation algorithms for polydisperse particles from nanometer to millimeter sizes without arbitrarily turning on or off specific force and torque terms. But, DHD is computationally intensive to the same degree as SD so that high-performance parallel computation is inevitable to simulate a reasonably large numbers of particles.

2.6. Discrete element method

In the dynamic motion of granular particles, ballistic collisions are one of the most fundamental and important interactions. Suppose two (spherical) particles i and j of mass m_i and m_j, respectively, undergo an inelastic collision, as shown in **Figure 1**. The relative velocity of particles i and j, denoted as g_{ij}, *at the point of contact* is determined by the translational and rotational particle velocities:

$$\boldsymbol{g}_{ij} = \boldsymbol{v}_{ij} - (a_i\boldsymbol{\omega}_i + a_j\boldsymbol{\omega}_j) \times \boldsymbol{n}_{ij} \tag{50}$$

where $v_{ij} = v_i - v_j$ is the relative velocity of the center of mass of particle i to j of velocity v_i and v_j, respectively. In general, particles are polydispersed, and a_i and a_j are the radii of particles i and j, respectively. The normal and tangential collision velocities given by the projections of g_{ij} are

$$\boldsymbol{g}_{ij}^n = (\boldsymbol{g}_{ij} \cdot \boldsymbol{n}_{ij})\boldsymbol{n}_{ij} \tag{51}$$

and

$$\boldsymbol{g}_{ij}^t = -\boldsymbol{n}_{ij} \times (\boldsymbol{n}_{ij} \times \boldsymbol{g}_{ij}) \tag{52}$$

respectively. The coefficients of restitution in normal and tangential direction, ϵ^n and ϵ^t, are defined as

$$(\boldsymbol{g}_{ij}^n)' = -\epsilon^n \boldsymbol{g}_{ij}^n \tag{53}$$

$$(\boldsymbol{g}_{ij}^t)' = +\epsilon^t \boldsymbol{g}_{ij}^t \tag{54}$$

where $0 \le \epsilon^n \le 1$ and $-1 \le \epsilon^{nt} \le 1$. The primed and unprimed variables indicate the pre- and post-collision quantities, respectively. Then, the velocities of particles i and j after the collision can be represented as functions of ϵ^n, ϵ^t, g_{ij}^n, g_{ij}^t, m_i, and m_j. Now, for particles i and j, we finally have

$$v'_i = \boldsymbol{v}_i - \left(\frac{\mu_{ij}}{m_i}\right)\left[(1+\epsilon^n)\boldsymbol{g}_{ij}^n + \frac{1-\epsilon^t}{1+1/\tilde{J}}\boldsymbol{g}_{ij}^t\right] \tag{55}$$

$$v'_j = \boldsymbol{v}_j + \left(\frac{\mu_{ij}}{m_j}\right)\left[(1+\epsilon^n)\boldsymbol{g}_{ij}^n + \frac{1-\epsilon^t}{1+1/\tilde{J}}\boldsymbol{g}_{ij}^t\right] \tag{56}$$

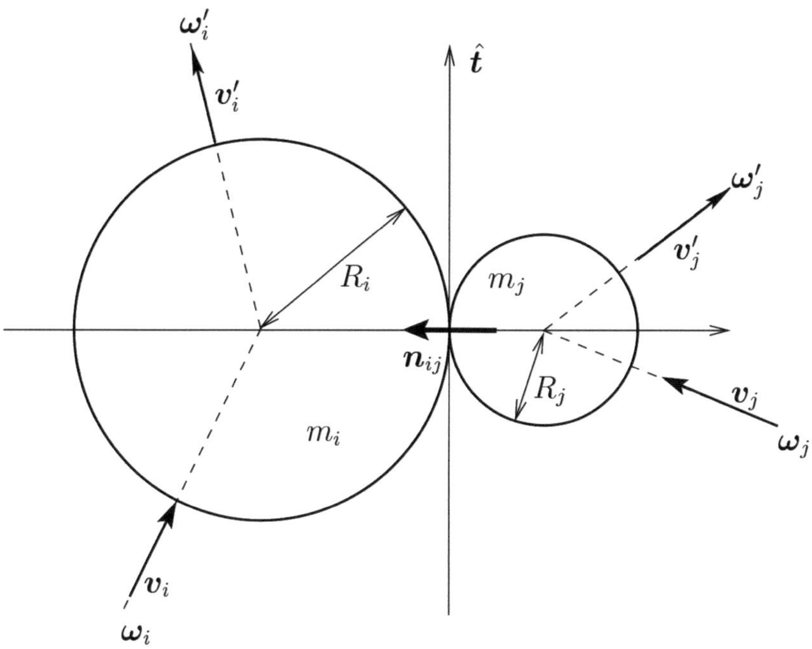

Figure 1. Collision of sphere *i* and *j*.

$$\omega'_i = \omega_i + \left(\frac{\mu_{ij}}{m_i a_i}\right)\left(\frac{1-\epsilon^t}{1+\tilde{J}}\right) \boldsymbol{n}_{ij} \times \boldsymbol{g}^t_{ij} \tag{57}$$

$$\omega'_j = \omega_j + \left(\frac{\mu_{ij}}{m_j a_j}\right)\left(\frac{1-\epsilon^t}{1+\tilde{J}}\right) \boldsymbol{n}_{ij} \times \boldsymbol{g}^t_{ij} \tag{58}$$

where for $k = i, j$. Here, J_k is the mass moment of inertia and $\tilde{J}_k = J_k/m_k a_k^2$ is its dimensionless form. A spherical particle has $J_k = 2/5$. The dimensionless mass moment of inertia for (rigid and spherical) particle is assumed to be constant. During the collision, the kinetic energy of particles is lost and transferred into thermal energy of the ambient fluid. At an arbitrary time before or after the collision, the kinetic energy of particle *k* is represented as

$$T_k^{(')} = \frac{1}{2} m_k \boldsymbol{v}_k^{(')} \cdot \boldsymbol{v}_k^{(')} + \frac{1}{2} J_k \boldsymbol{\omega}_k^{(')} \cdot \boldsymbol{\omega}_k^{(')} \tag{59}$$

One can calculate the energy loss of particle *k*, $\Delta T_k = T'_k - T_k$, using the pre- and post-collision velocities of $(\boldsymbol{v}_k, \boldsymbol{\omega}_k)$ and $(\boldsymbol{v}'_k, \boldsymbol{\omega}'_k)$, respectively. Post-collision velocities of two unequal spheres are completely solved using Eqs. (**55**)–(**58**), but an underlying assumption is that the inelastic collision is instantaneous without spending any time. In reality, however, any granular collision takes a finite amount of time, even if it is much shorter than the traveling time of the particles. This time duration in which two particles are in contact is defined as *contact time* (or *collision duration*). Therefore, it is more accurate to see the collision of two granules as an impulse event. The collision rule well determines the post-collision states, if the granules are in a fluid-like state. But, when densely-packed granules are slowly moving under mechanical or gravitational compression, the collision rule fails to predict the transient granular states because it does not take into account the compression and restoration of the inelastic particles.

2.7. Overview

Table 1 summarizes specific features of the above-discussed simulation methods. Acceleration can be included by all methods in principle, but is barely employed in BD and SD. This is mainly due to the time interval restriction. Instead, the fluctuation-dissipation theorem is intrinsically embedded and satisfied in BD, SD, DPD, and DHD formalisms. All the simulation methods can surely include effects of conservative, external force fields. Brownian motion and multi-body hydrodynamic forces are (most) accurately implemented in SD and DHD, but only hydrodynamics is important in the many-body motion of non-Brownian granules. The DHD is the only simulation method that can mimic many-body hydrodynamics without the time interval restriction. Constraint forces/torques can be easily applied to any dynamics methods to simulate compound particles or aggregates. SD has the same capability but the collision rules are not included. Due to typical ranges of particle sizes in the simulation methods, the inelastic collision (or contact) forces are included only in DEM. Since DPD allows particle deformation by using the relaxed, pairwise hydrodynamic resistance between two particles, i.e., ω_D and ω_R in Eq. (44), employing the collision force of rigid bodies in DPD must be out of its original scope. The current state-of-the-art DEM algorithms can be further improved to mimic complex phenomena of dry granules of various shapes [50, 51]. However, in our opinion, granular motion in a liquid phase can be accurately simulated by only using DHD or SD. To accurately simulate a large number of granules of various sizes in complex fluid environments, multi-body hydrodynamics must be rigorously implemented in the current DEM method.

		MD	BD	SD	DPD	DHD [44]	DEM
Gov. Eq.		$F = ma$	Langevin	Langevin	SDE	**SDE**	various
Acceleration		Yes	Possible	Possible	Yes	**Yes**	Yes
Force	Conservative	Yes	Yes	Yes	Yes	**Yes**	Yes
	Brownian	No	Simple	Rigorous	Approx.	**Rigorous**	No need
	Hydrodynamics	No	Constant	Accurate	Pairwise	**Accurate**	No
	Constraint	Yes	Yes	Yes	Yes	**Yes** [54]	Yes
	Collision	No	No	No	No	**(possible)**	Yes
Time interval dt		Any	$\delta t \gg \tau$	$\delta t \gg \tau$	Any	**Any**	Any

Table 1. Comparison of particle dynamics simulation methods. The conservative force includes external forces such as electromagnetic and gravitational, and inter-particle force such as DLVO [52, 53]. Here, "any" dt indicates that dt can be arbitrarily determined for accurate and fast calculation regardless of the particle relaxation time τ.

3. Granular dynamics: theory and simulations

Granular materials consist of a large number of particles whose typical size ranges from micrometers to centimeters [50, 51]. These particles interact via short-range forces through only mechanical contacts and (external) long-ranged electromagnetic or gravitational forces. Granular dynamics mimics dynamic motion of granular particles in a transient state such as excited or granules in a fluid media. Large-scale phenomena in this category include land sliding and snow avalanches. Wet granules such as sand in beaches undergo hydrodynamic forces due to tidal currents. Soil granules in unsaturated sub-surfaces are partially dry or wet. Interstitial water layers between granules can significantly change inter-granular interactions, especially when they are in a stationary contact phase. The significance of hydrodynamic interactions between non-Brownian granules is paid less attention. In this section, we describe granular dynamics as a microscopic extension of DEM in a shorter time scale by investigating the collision phenomenon between two (spherical) particles, as shown in **Figure 1**, during an impulse event.

3.1. Individual contact forces

3.1.1. Normal and tangential forces

Small particles such as suspended solid, colloids, and nanoparticles are naturally negatively charged. Their electrostatic repulsive forces decay exponentially with respect to the distance from their surfaces to the bulk phase. When the surface-to-surface distance between two particles is much smaller than their average sizes, forces such as electrostatic and Born repulsion are strong enough to repel each other. These forces are, however, not dominant for large, non-Brownian particles such as granules of an order of $O(0.1 - 10)$ mm. The dominant granular force stems from contact during collisions. In conventional statistical mechanics, a hard sphere is characterized by the wall potential:

$$V = \begin{cases} 0 & r > 2a \\ \infty & r < 2a \end{cases} \tag{60}$$

where a is the particle radius. The mathematical discontinuity of the wall potential at $r = 2a$ indicates the infinite force that completely prevents any overlap between particles.

As granules are inelastic, the fundamental wall potential must be modified before it is applied to a granular system. Two spherical granules are in a mechanical contact if the sum of their radii exceeds their center-to-center distance, i.e.,

$$\xi_{ij} \equiv a_i + a_j - |\boldsymbol{r}_i - \boldsymbol{r}_j| > 0 \tag{61}$$

where ξ_{ij} is the *mutual compression* of particles i and j. Note that ξ_{ij} is positive when two granules overlap and becomes larger when the granules come closer. Thus, the mutual compression can be interpreted as the overlapped surface-to-surface distance. The force acting on particle i from particle j (conceptually denoted as $i \leftarrow j$) is described by

$$\boldsymbol{F}_{ij} = \begin{cases} \boldsymbol{F}^{n}_{ij} + \boldsymbol{F}^{t}_{ij} & \text{for} \quad \xi_{ij} > 0 \\ 0 & \text{for} \quad \xi_{ij} \leq 0 \end{cases} \tag{62}$$

where $\boldsymbol{F}^{n}_{ij}$ and $\boldsymbol{F}^{t}_{ij}$ are the normal and tangential components of the contact force, respectively. For simplicity, Eq. (62) can be rewritten as

$$\boldsymbol{F}_{ij} = (\boldsymbol{F}^{n}_{ij} + \boldsymbol{F}^{t}_{ij})H(\xi_{ij}) \tag{63}$$

where $H\,(x)$ is the Heaviside step function, defined as

$$H(x) = \begin{cases} 1 & \text{for} \quad x > 0 \\ 0 & \text{othersies} \end{cases} \tag{64}$$

In three-dimensional space, vector quantities between particles i and j can be decomposed into the normal and tangential directions. Using the mathematical identity

$$\boldsymbol{A} \times (\boldsymbol{B} \times \boldsymbol{C}) = \boldsymbol{B}(\boldsymbol{A} \cdot \boldsymbol{C}) - \boldsymbol{C}(\boldsymbol{A} \cdot \boldsymbol{B}) \tag{65}$$

one can replace $\boldsymbol{A}$ and $\boldsymbol{B}$ by $\boldsymbol{n}$ and $\boldsymbol{C}$ by $\boldsymbol{F}_{ij}$ to write

$$\boldsymbol{n} \times (\boldsymbol{n} \times \boldsymbol{F}_{ij}) = \boldsymbol{n}(\boldsymbol{n} \cdot \boldsymbol{F}_{ij}) - \boldsymbol{F}_{ij} \tag{66}$$

$$\boldsymbol{F}_{ij} = \boldsymbol{n}(\boldsymbol{n} \cdot \boldsymbol{F}_{ij}) - \boldsymbol{n} \times (\boldsymbol{n} \times \boldsymbol{F}_{ij}) \tag{67}$$

From Eq. (67), the normal and tangential force components can be expressed as

$$\boldsymbol{F}^{n}_{ij} = (\boldsymbol{n} \cdot \boldsymbol{F}_{ij})\boldsymbol{n} \tag{68}$$

$$\boldsymbol{F}^{t}_{ij} = (\boldsymbol{n} \times \boldsymbol{F}_{ij}) \times \boldsymbol{n} \tag{69}$$

The contact force between elastic spheres was originally developed by Hertz [55] and was later generalized for viscoelastic (damped) particles [56, 57] as

$$F^{n} = \frac{2Y\sqrt{a_{\text{eff}}}}{3(1-\nu^{2})}\left(\xi^{3/2} + A\sqrt{\xi}\frac{d\xi}{dt}\right) \tag{70}$$

where Y and ν are Young's modulus and Poisson's ratio, respectively, A is the dissipative constant being a function of material viscosity, and a_{eff} is the effective radius that can be interpreted as a harmonic sum a_i and a_j, i.e., $a_{\text{eff}}^{-1} = a_i^{-1} + a_j^{-1}$. Parameter A explains the dependence of the restitution coefficient on the approaching velocity between two spheres. If $A = 0$, then Eq. (70) converges to the original Hertz's equation for elastic granules. Therefore, parameter A needs to be inversely calculated using an experimentally observed coefficient of restitution. If two elastic particles are heterogeneous, then Hertz's equation may be extended to

$$F_{ij}^{n} = \frac{4\sqrt{a_{\text{eff}}}}{3}\left(\frac{1-\nu_i^2}{Y_i}+\frac{1-\nu_j^2}{Y_j}\right)^{-1}\xi^{3/2} \tag{71}$$

for particles of different Y and ν values. Assuming the dissipative constant A is also particle-specific and additive, the most general expression of the normal force between two viscoelastic spheres may be [51]

$$F_{ij}^{n} = \frac{4\sqrt{a_{\text{eff}}}}{3}\left(\frac{1-\nu_i^2}{Y_i}+\frac{1-\nu_j^2}{Y_j}\right)^{-1}\left(\xi^{3/2}+\frac{A_i+A_j}{2}\dot{\xi}\sqrt{\xi}\right) \tag{72}$$

as a generalized extension from Eq. (70). Note that this viscoelastic force is finite while two spheres are being overlapped so that the mutual compression ξ is positive. Let's define $t = 0$ as the moment of the contact of two particles. The compression continues until $t = t_c$, after which restoration begins. When the relative velocities between the two particles at $t = 0$ and $t = t_c$ are $\dot{\xi}(0)$ and $\dot{\xi}(t_c)$, respectively, then the normal restitution coefficient can be calculated by measuring velocities $\dot{\xi}(0)$ and $\dot{\xi}(t_c)$, i.e.,

$$\epsilon^n = \dot{\xi}(t_c)/\dot{\xi}(0) \tag{73}$$

This indicates that ϵ^n depends on the relative collision velocity, unless $A = 0$. Theoretically, at least three experiments are required for collisions between particles i and j for pairs of (i, j), (i, i), and (j, j). Then, A_i and A_j can be inversely calculated by using the trial-and-error method for numerical fitting.

The relative velocity of the spheres at the point of contact results from the relative translational/rotational velocities. The contact-point velocity has the tangential component of

$$\boldsymbol{v}_{\text{rel}}^{t} = (\boldsymbol{v}_j - \boldsymbol{v}_i)\cdot \boldsymbol{e}_{ij}^{t} + a_i\boldsymbol{\omega}_i + a_j\boldsymbol{\omega}_j \tag{74}$$

which provides the tangential force of

$$F^t = -\text{sign}(\boldsymbol{v}_{\text{rel}}^{t})\cdot \min(\gamma^t|\boldsymbol{v}_{\text{rel}}^{t}|, \mu|F^n|) \tag{75}$$

where γ^t is a fitting coefficient, proportional to the tangential dissipative force in magnitude. In Eq. (75), the shear force is limited by Coulomb's friction law of $|F^t| \le \mu|F^n|$, where μ is the friction coefficient. Although this approach is conceptually straightforward, the level of approximation is still on the collision rule, as discussed in Section 2.6. Unless F^n is constant, Eq. (75) provides an inconsistent value of F^t because F^n (if Eq. (72) is used) is a function of not only the mutual compression, but also the relative velocity. The tangential contact force is correlated to the normal force [56, 57]:

$$F_{ij}^{t} = -\mu F_{ij}^{n}\left(\frac{\zeta_{ij}}{\zeta_0} - \left\lfloor\frac{\zeta_{ij}}{\zeta_0}\right\rfloor\right) \tag{76}$$

where ζ is the relative tangential shift, ζ_0 is its macroscopic maximum value, and $\lfloor x \rfloor$ denotes the integer of x. Typical values of ζ_0 / a_{eff} range from 10^{-7} to 10^{-3}. Similar to Eq. (73), the tangential coefficient of restitution can be calculated as

$$\epsilon^t = \dot{\zeta}(t_c)/\dot{\zeta}(0) \tag{77}$$

If the friction coefficient μ is known, experimental measurement of ϵ^t allows us to inversely calculate ζ_0. Techniques to determine the coefficients of restitution of colliding viscoelastic spheres can be found elsewhere [58]. This approach can be readily applied to densely packed granular medium with small movements or vibrations such as the Brazilian bean problem [59–64]. However, time integration per collision, consisting of compression and restoration, should be from 0 to t_c, while t_c is a negligibly short time in the collision-rule approach. For efficient computations, regular time-integrations and event-based simulations should be efficiently combined in order to have variable δt, which should be much smaller than t_c in compressing/ restoring phases, and much longer than t_c for collision events for fluidized granules of a low concentration.

3.1.2. Shear stress

Shear thickening: When non-Brownian granules are densely packed, volume fraction is around 50%, depending on their polydispersity and shape. The granules form a loosely connected material, which responds to the external shear stresses in an unconventional way. Depending on the magnitude of the external stress, granules temporarily switch their phases between the liquid-like and the solid-like states. In a (pure) fluid, viscosity is defined as the ratio of shear stress to the shear rate during a steady flow, which represents the energy dissipation rate by the fluid flow. This dissipation rate in some cases decreases with respect to the shear rate, which is known as *shear thinning*. It is particularly desirable for paints such that pigments flow easily when brushed, but does not drip when brushing stops. On the other hand, *shear thickening* indicates that the energy dissipation rate increases as the shear rate increases. In other words, the fluid becomes much more viscous if the external stress is strong enough. For example, if a large amount of cornstarch or Oobleck is mixed with water in a (small) swimming pool [65, 66],[1] the dense suspension acts like a liquid if it is at rest in the absence of external stresses applied. But, when the suspension is sheared, the flow resistance increases dramatically and the fluid becomes locally amorphous for a short amount of time. If a person is continuously stepping on the dense suspension, the person will be able to dynamically stay on top of the semi-liquid surface. When the person slows down or stops moving, then he or she will slowly sink into the pool.[2] This phenomenon is not only interesting, but also has practical importance to engineering systems like automobile brakes [67].

Possible mechanisms of this shear-thickening include hydroclustering, order-disorder transition, and dilatancy. First, in hydroclustering, particles gather together into transient, reversible clusters under shear flow, and this rearrangement leads to increases in the lubrication drag forces [68, 69]. Local heterogeneity of particle suspension creates regions in which particles undergo less drag forces and tend to agglomerate. This results in narrow flow channels among the dynamic groups of particles. Application of Stokesian dynamics to mimic the hydroclustering intrinsically prevents

[1] A fictional green substance in the Dr. Seuss book Bartholomew and the Oobleck.

[2] Can You Walk on Water? (Non-Newtonian Fluid Pool) https://youtu.be/D-wxnID2q4A

particle contacts, followed by inelastic overlaps in a series of collision events. This is because the lubrication force is logarithmically proportional to the surface-to-surface distance between two particles and thus it diverges when two particles start inelastically overlapping. Second, the order-disorder phase transition includes the changes in the flow structure between ordered and disordered phases, which yields increases in the drag force between particles [70]. These ordered states are different from fixed 3D structures found in solids, but similar to amorphous aggregates. Third, the dilatancy mechanism describes the shear-thickening such that the effective volume of particle packing increases under the shear. When particles are confined in local spaces or partially jammed, the shear force pushes particles toward the containing walls. Additional stress can be developed on the walls and backward responses may generate extra stress between particles and walls [71] in contact interfaces. Recently, discontinuous shear thickening (DST) was proposed to explain the shear thickening experiments [69], of which comprehensive review can be found elsewhere [72, 73]. To the best of our knowledge, the fundamental mechanism of the shear thickening has not been fully discovered.

Sample simulation: Figure 2 shows results of a sample simulation using the mechanisms described in Section 3.1.1, which can be further extended to DST simulations. The gray spheres are loosely packed, ideally forming a body-centered cubic structure. The top layer consists of 5 × 5 = 25 regularly packed spheres, under which 6 × 6 = 36 spheres are located. These spheres

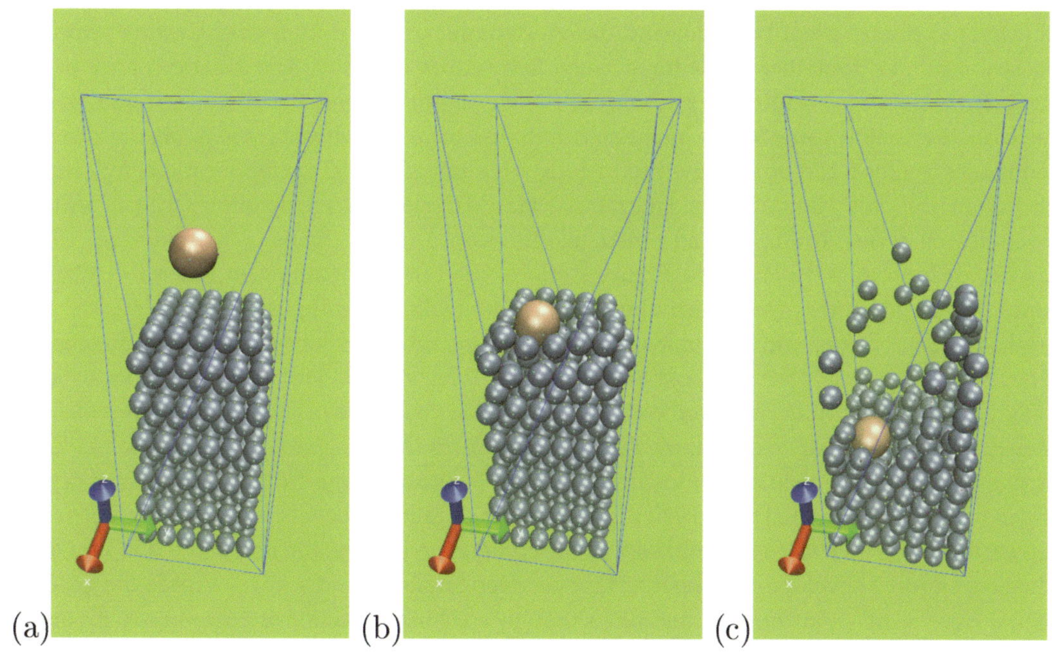

Figure 2. Simulation of granular damping to a sudden impact: an intruder (a) approaches with high speed, (b) collides with a few granules on the packed surface, and (c) penetrates the loosely packed inelastic (energy absorbing) granules. Snapshots are visualized using Visual Molecular Dynamics (VMD) [74–76]. Radii of the intruder and packed spheres are 2 and 1 mm, respectively, and their specific gravity values are commonly 2.75. The spheres and the walls have Poisson's ratio of 0.4 and 0.6, and Young's modulus of 4.0×10^7 Pa and 1.0×10^9 Pa, respectively. The introduced energy dissipation rate A was set at 2.5×10^{-5}, for both spheres and walls [56].

are closely located to each other but not touching, having very small surface-to-surface distance between the nearest neighbors. The two layers are packed five times vertically, and therefore the total number of gray particles is 366. This preliminary simulation aims to see the impact responses of the small packed spheres when hit by a big, fast intruder. The intruder particle initially moves with the downward velocity of $v_z = 13.00 \times 10^{-3}$ m/s and spins with angular velocity of $\omega_z = 0.56$ rad/s.

Due to the loose packing of the small spheres on the bottom of the container, the intruder tends to penetrate the packed granules. In the initial stage of the penetration, the intruder is surrounded by small spheres, which bounce away from their initial positions. If the small spheres are initially touching each other, the force propagation from the top layer to the bottom layer must be almost instantaneous. The intruder will experience a strong normal force and may rebound in the opposite direction after an initial penetration of a short depth. As the collective phenomena of many granules are mostly transient, not only the material properties of the intruder but also initial and boundary conditions of granules play significant roles in their macroscopic dynamic behaviors. High dissipation rate of A reduces the impact of the one-to-many collisions between the intruder and dissipating granules. The intruder's impact is transmitted through dynamic chains of contacting spheres, but not all of the packed granules participate in the force transmission. There are same granules almost free from the intruder's impact.

Calculation of forces and torques exerted on each particle and their visualization can help understand the transient behavior and design dynamic granular materials. Granules initially located near the container walls have much less spaces to move. The kinetic energy of the intruder is transmitted to these granules at the cul-de-sac and mostly dissipated on the wall surfaces. Returning force to the intruder is initiated from the boundary. Applications of this simulation method can be designed to include a large number of real applications for characterization and prediction of transient granular material properties. One important key issue is, as indicated above, to identify and visualize the transmission chains of force/torque, which dynamically form and disappear. **Figure 3** shows the initial impact event when the intruder starts penetrating the loose granular packing, visualized using open visualization tool (OVITO) [77]. The top and bottom rows show the top and side views of the intruder collision. The left column shows particle positions as well as force vectors, and the right column shows only particle configurations where color indicates the magnitude of the net force. Impact from the intruder is somehow irregularly distributed around the top granules. On the left-column images, arrows and their colors indicate force directions and magnitudes. A number of downward vectors indicate that the gravitational force is dominant for non-touching granules. The upward arrow in the intruder implies that the contact force to the intruder, which is similar to the normal force developed on the interface between an object and a wall, exceeds the gravitational force exerted on the intruder. Even though the force direction is temporarily inverted, the intruder still goes down due to the inertia of the high initial velocity. **Figure 3** clearly shows that only a partial number of granules participate in the force transmission from the intruder to the packed granules. These force chains are very transient, and more importantly, the magnitude of the transmitted force diminishes as time goes by due to the intrinsic inelastic nature of the granules. For the future, designs of smart damping materials can be achieved not only by understanding various mechanical properties of the granules, but also by controlling specific

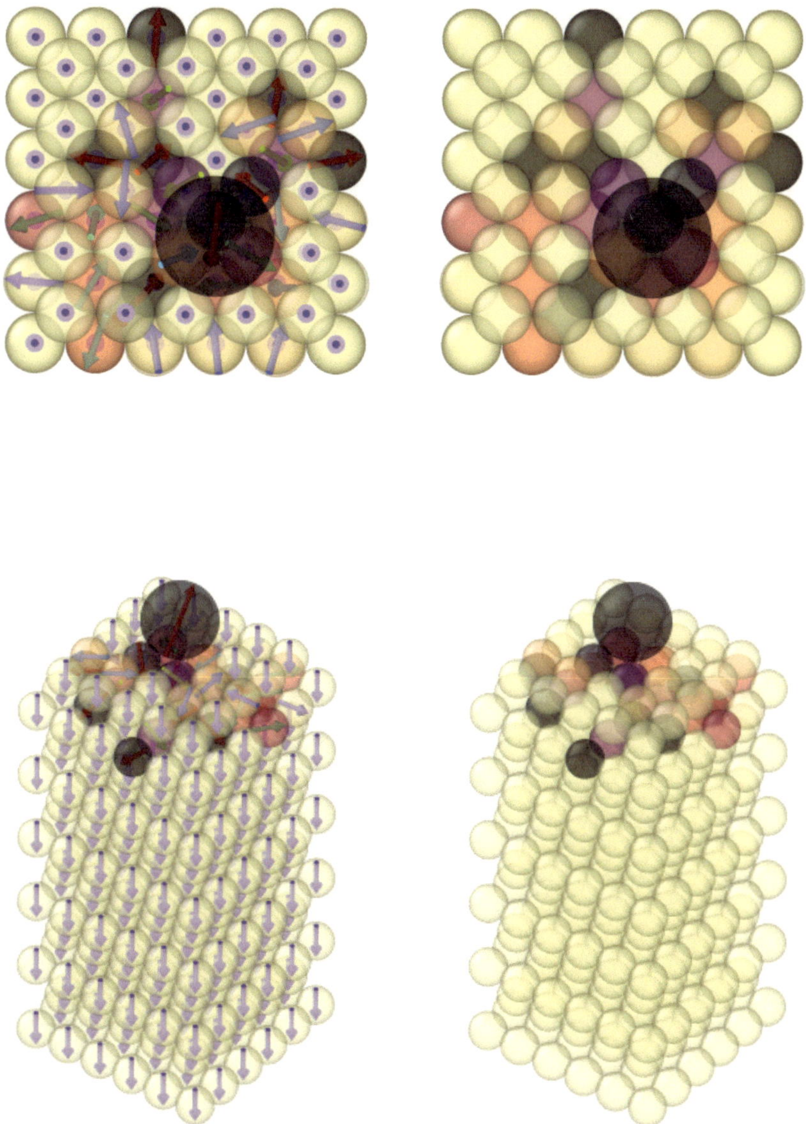

Figure 3. Force chain visualization of the intruder using open visualization tool (OVITO).

initial and boundary condition, which leads inelastic many-granule systems to behave as smart materials.

3.2. Constraint force: holonomic and non-holonomic

3.2.1. Holonomic potential for translational constraints

Irregular-shaped granules can be mimicked as a large collection of polydispersed spherical particles. For simplicity, we first consider two particles (point masses) moving together as one body. A constraint embedded between the two particles keeps the inter-particle distance invariant. A translational constraint between particle i and j is

$$\sigma_{ij} = (\boldsymbol{r}_i - \boldsymbol{r}_j)^2 - d_{ij}^2 = 0 \tag{78}$$

where d_{ij} is the fixed distance between particles i and j, usually an average of two contacting rigid bodies of diameters d_i and d_j, i.e., $d_{ij} = (d_i + d_j)/2$. This type of geometrical constraint is called *holonomic*. To develop an inter-particle interaction to satisfy Eq. (78), one defines the constraint potential for all N_p particles as

$$\Phi = \frac{1}{2}\sum_{i=1}^{N_p}\sum_{j>i}^{N_p}\lambda_{ji}\sigma_{ij} \tag{79}$$

where λ_{ji} is a symmetric Lagrange's multiplier and $\frac{1}{2}$ in the front of the summation is by convention. Exchanging positions of particles i and j ($i \leftrightarrow j$) should not change the sign and the magnitude of Φ. To satisfy this condition, the Lagrange multiplier is symmetric, i.e., $\lambda_{ij} = \lambda_{ji}$. The constraint force exerted on particle j can be derived as a negative derivative of the constraint potential:

$$\boldsymbol{F}_j^C = -\nabla_j\Phi = \sum_{i\neq j}\lambda_{ji}(\boldsymbol{r}_i - \boldsymbol{r}_j) \tag{80}$$

where i runs from 1 to N_p except for the case $i = j$ (if so, $\lambda_{ii} = 0$), or simply

$$\boldsymbol{F}_j^C = \sum_{i=1}^{N_p}\lambda_{ji}(\boldsymbol{r}_i - \boldsymbol{r}_j)(1 - \delta_{ij}) \tag{81}$$

where δ_{ij} is the Kronecker delta symbol, defined as

$$\delta_{ij} = \begin{cases} 1 & \text{for} \quad i = j \\ 0 & \text{otherwise} \end{cases} \tag{82}$$

Among $N_p - 1$ pairs made by particle j (excluding itself), if particle j does not have any constraint to particle k, then $\lambda_{jk} = 0$, and symmetrically vice versa. For a two-body case, the holonomic constraint force acting on j by k is

$$\boldsymbol{F}_{j\leftarrow k}^C = -\nabla_j\Phi = \lambda_{jk}(\boldsymbol{r}_j - \boldsymbol{r}_k) \tag{83}$$

and, similarly, the same force on k by j is

$$\boldsymbol{F}_{k\leftarrow j}^C = -\nabla_k\Phi = \lambda_{kj}(\boldsymbol{r}_k - \boldsymbol{r}_j) \tag{84}$$

Since λ_{jk} is symmetric ($\lambda_{jk} = \lambda_{kj}$), one can show that

$$\boldsymbol{F}_{j\leftarrow k}^C = -\boldsymbol{F}_{k\leftarrow j}^C \tag{85}$$

which follows Newton's third law, *the action and reaction* principle. Summation over all constraint particles will give a zero resultant force. This is because the constraint force is an internal force and the sum of internal forces is zero.

Evolution of positions: Translational acceleration of particle j can be expressed as

$$\boldsymbol{a}_j = \boldsymbol{a}_j^{\dagger} + \boldsymbol{a}_j^{C} \tag{86}$$

where $a_j^{\dagger}$ is the unconstrained acceleration and a_j^{C} is the constrained acceleration, represented as

$$\boldsymbol{a}_j^{C} = \frac{1}{m_j}\lambda_{jk}(\boldsymbol{r}_k - \boldsymbol{r}_j) \tag{87}$$

Assume particle j evolves from $\boldsymbol{r}_j(t)$ at time t to $\boldsymbol{r}(t+\delta t)$ at time $t+\delta t$. Then, the future position at $t+\delta t$ can be decomposed into two parts:

$$\boldsymbol{r}_j(t+\delta t) = \boldsymbol{r}_j^{\dagger}(t+\delta t) + \frac{1}{2}\boldsymbol{a}_j^{C}\delta t^2 \tag{88}$$

where

$$\boldsymbol{r}_j^{\dagger}(t+\delta t) = \boldsymbol{r}_j(t) + \boldsymbol{v}_j(t)\delta t + \frac{1}{2}\boldsymbol{a}_j^{\dagger}\delta t^2 \tag{89}$$

is the evolved position at time $t+\delta t$ in the absence of the constraint force. A similar equation for particle k can be written easily. Note that Eq. (78) should be valid at all times. Substitution of Eq. (88) into (78) gives, neglecting terms on the order of (δt^4) and higher, the representation of the Lagrange multiplier:

$$\lambda_{ij}\delta t^2 = \frac{[\Delta\boldsymbol{r}_{ij}^{\dagger}(t+\delta t)]^2 - d_{ij}^2}{\mu_{ij}^{-1}[\Delta\boldsymbol{r}_{ij}^{\dagger}(t+\delta t)]\cdot[\Delta\boldsymbol{r}_{ij}(t)]} \tag{90}$$

where

$$\Delta\boldsymbol{r}_{ij}^{\dagger}(t+\delta t) = \boldsymbol{r}_i^{\dagger}(t+\delta t) - \boldsymbol{r}_j^{\dagger}(t+\delta t) \tag{91}$$

$$\Delta\boldsymbol{r}_{ij}(t) = \boldsymbol{r}_i(t) - \boldsymbol{r}_j(t) \tag{92}$$

and

$$\mu_{ij}^{-1} = m_i^{-1} + m_j^{-1} \tag{93}$$

is called the reduced mass of particles i and j. In Eq. (90), λ_{ij} can be determined using an iterative method.

1. Initially, the unconstrained position $\boldsymbol{r}_j^{\dagger}(t+\delta t)$ is calculated using Eq. (89).

2. Insert $\boldsymbol{r}^{\dagger}(t+\delta t)$ into Eq. (90) to calculate λ_{ij}.

3. Calculate the constraint acceleration of particle j, $\boldsymbol{a}_j^{C}$, using Eq. (87).

4. Update the particle position using Eq. (88), which is under the influence of constrained and unconstrained accelerations. Set this updated position as the unconstrained position at that time: $\boldsymbol{r}_j^\dagger(t+\delta t) \leftarrow \boldsymbol{r}_j(t+\delta t)$.

5. Having the updated $\boldsymbol{r}_j^\dagger$, go to step 2, unless λ_{ij} converges to a finite value. Otherwise, store information at $t + \delta t$, and go to the next time step.

This iterative procedure will continue until the Lagrange multiplier converges within a preset tolerable error for the position evolution from time t to $t + \delta t$. So far, the constraint force modifies the particle position at time $t + \delta t$ from its unconstrained position $\boldsymbol{r}_j^\dagger(t+\delta t)$, but the velocity after the constrained evolution of position is the same as before the evolution.

Velocity evolution: Differentiation of the holonomic constraint Eq. (78) with respect to time gives

$$(\boldsymbol{r}_i - \boldsymbol{r}_j) \cdot (\boldsymbol{v}_i - \boldsymbol{v}_j) = 0 \tag{94}$$

valid both at time t and $t + \delta t$. Similar to the position evolution, the translational velocity at time $t + \delta t$ is represented as

$$\boldsymbol{v}_j(t') = \boldsymbol{v}_j^\dagger(t') + \boldsymbol{a}_j^C \delta t \tag{95}$$

where

$$\boldsymbol{v}_j^\dagger(t') = \boldsymbol{v}_j(t) + \boldsymbol{a}_j^\dagger \delta t \tag{96}$$

is the updated velocity from time t without the holonomic constraint. Substitution of Eq. (95) into Eq. (94) gives

$$\Delta\boldsymbol{r}_{ij}(t+\delta t) \cdot \Delta\boldsymbol{v}_{ij}^\dagger = -\Delta\boldsymbol{r}_{ij}(t+\delta t) \cdot \Delta\boldsymbol{a}_{ij}^C \delta t \tag{97}$$

where

$$\Delta\boldsymbol{a}_{ij}^C = \frac{\kappa_{ij}}{\mu_{ij}}(\boldsymbol{r}_j - \boldsymbol{r}_i) \tag{98}$$

is the constraint acceleration for velocity correction. Here, κ_{ij} plays a similar role of λ_{ij}, but it is independently determined only to update the velocity, which is calculated as

$$\kappa_{ij}\delta t = \frac{\Delta\boldsymbol{r}_{ij}(t+\delta t) \cdot \Delta\boldsymbol{v}_{ij}^\dagger}{d_{ij}^2 \mu_{ij}^{-1}} \tag{99}$$

To update κ_{ij}, a similar iteration method can be used:

1. Calculate the unconstrained velocity at the next time step, $v^\dagger$ $(t + \delta t)$ for particles i and j, and calculate their difference $\Delta\boldsymbol{v}_{ij}^\dagger = \boldsymbol{v}_i^\dagger - \boldsymbol{v}_j^\dagger$ at time $t + \delta t$.

2. Calculate κ_{ij} using Eq. (99) using previously determined $\boldsymbol{r}_i(t + \delta t)$ and $\boldsymbol{r}_j(t + \delta t)$.

3. Update $\Delta \boldsymbol{a}_{ij}^C$ using Eq. (98) and use it to calculate $\boldsymbol{v}_j\ (t + \delta t)$ of Eq. (95).

4. Replace $\boldsymbol{v}_j\ (t + \delta t)$ by $\boldsymbol{v}_j^\dagger(t + \delta t)$ and go to step 2 unless κ_{ij} converges to a constant value within a tolerable error. Otherwise, store information at $t + \delta t$, and go to the next time step.

So far, we first defined the constraint potential using unknown Lagrange's multipliers, derived the constraint force and acceleration, and updated iteratively particle positions and velocities until all the constraints are satisfied. The velocity evolution under this constraint is similarly done by introducing a new independent Lagrange's multiplier, which is to satisfy the orthogonal relationship between position and velocity variations, in Eq. (94).

Figure 4 shows a simple test of the holonomic constraint between two unequally-sized spheres. It is clear that the two spheres are in contact with each other during their translational motion. A camera is moving with the same velocity of their center of mass so that only the relative motion is shown. Both the particles are non-Brownian, and the red sphere is 1.5 times bigger than the blue one, while their specific gravity is 2.75. The blue sphere rotates in the clock-wise direction around the red sphere. This is because the red one is $1.5^3 = 3.375$ times heavier so that the total center of mass is closer to that of the red particle. Careful observation indicates that the blue sphere rotates in the clock-wise direction about its center of mass, and the red sphere rotates in the opposite direction about its center of mass. This relative rotation stems from the presence of only holonomic (translational) constraints, which allows their smooth surface-elements to slide relative to each other. This phenomenon must happen if weakly attractive particles form loose aggregates in a fast viscous flow.

3.2.2. Non-holonomic torque for angular constraints

Dynamics simulations with the holonomic constraints work perfectly within tolerable errors, especially when the particles sizes are smaller than center-to-center distances. This method was successfully used for molecules of fixed structures such as water (H_2O) and organic compounds. On the other hand, if two spherical particles (such as colloids) of finite volumes are attached by sticky surface forces, the translational and rotational motion of these two spheres are constrained as they move as a single compound body. For simplicity, we will consider a pair of compounded golf balls. If only the two constraints discussed in Section 3.2.1 are considered regarding the motion of the compounded golf balls, their rotations about their own centers of mass are still allowed even if surface friction exists. Mathematically, the two balls, if perfectly glued on their small shared surfaces, should have the same angular velocity. This type of constraint is based on (angular) velocity so it is called *non-holonomic.*

Figure 4. A dimer of bi-dispersed spheres of holonomic constraint only.

Consider a perfectly inelastic collision event between two approaching particles i and j, moving with translational and rotational velocities of (v_i, ω_i) and (v_j, ω_j), respectively, before their collision. For $k = i, j$, the linear and angular momenta are

$$\boldsymbol{p}_k = m_k \boldsymbol{v}_k \tag{100}$$

$$\boldsymbol{H}_k = J_k \boldsymbol{\omega}_k \tag{101}$$

respectively. After the two particles are permanently attached, the total linear momentum is

$$M_a \boldsymbol{V}_a = m_1 \boldsymbol{v}_1 + m_2 \boldsymbol{v}_2 \tag{102}$$

where $M_a = m_1 + m_2$ is the total mass of the two-body aggregate and $\boldsymbol{V}_a$ is the translational velocity of the center of mass of the aggregate. The total angular momentum has, however, a slightly different formulation:

$$\boldsymbol{H}_a = (J_1 + m_2 d_1^2)\boldsymbol{\omega}_1 + (J_2 + m_2 d_2^2)\boldsymbol{\omega}_2 = J_a \Omega_a \tag{103}$$

where

$$J_a = I_1 + m_1 d_1^2 + I_2 + m_2 d_2^2 \tag{104}$$

is the mass moment inertia of the aggregate. Here, d_k is the shortest distance between the center of mass of particle k and the rotation axis of the aggregate, passing through the center of mass of the aggregate, $\boldsymbol{r}_{CM}$:

$$d_k = \frac{(\boldsymbol{r}_k - \boldsymbol{r}_{CM}) \cdot \Omega_a}{|\Omega_a|} \tag{105}$$

After the perfectly inelastic collision, two compounded particles will have the same angular velocity $\boldsymbol{\Omega}_a$. The total torque acting on the aggregate is rigorously represented as

$$\boldsymbol{M} = \sum_j (\boldsymbol{r}_j - \boldsymbol{R}_{CM}) \times (\boldsymbol{F}_j^{\mathrm{Ex}} + \boldsymbol{F}_j^{\mathrm{C}}) \tag{106}$$

and the time evolution equations related to the total angular momentum are

$$\frac{d\boldsymbol{H}_a}{dt} = \boldsymbol{M} \tag{107}$$

$$\boldsymbol{H}_a(t + \delta t) = \boldsymbol{H}_a(t) + \boldsymbol{M}\delta t \tag{108}$$

$$\Omega_a(t + \delta t) = \Omega_a(t) + J_a^{-1} \boldsymbol{M} \delta t \tag{109}$$

Finally, all associated particles to the aggregate have the same angular velocity $\boldsymbol{\Omega}_a$ at any time unless they break apart of slide into each other. It is assumed that the mutual compressions ξ

between the associated particles are small enough to be neglected in calculating the aggregate's mass moment of inertia, J_a.

Figure 5 compares dynamics of a trimmer, i.e., three constrained bodies. The six small black particles per sphere are imaginary, showing how the particle rotates in its transient motion. These imaginary markers do not generate any forces or torques. Initially, three spheres associated with a trimmer make the ideal L-shape. The downward gravitational force causes the settling of the trimmer. As noted above, the camera is moving with exactly the same velocity of the trimmer's center of mass. Therefore, one can observe only relative motion of three identical spheres with respect to their center of mass. The gap between the two closest spheres is equal to the diameter of the black marker. In **Figure 5(a)**, three particles undergo only holonomic constraints such that the center-to-center distances are kept constant. As the outside surfaces of each particle experience higher hydrodynamic stress, all three particles try to rotate toward the center, as viewed from the top of the trimmer. On the other hand, **Figure 5(b)** shows a trimmer of three rigidly attached (glued) spheres. Relative rotation of a sphere with respect to its neighbors is prevented. This indicates that all the three spheres in **Figure 5(b)** have identical angular velocity, which is equal to that of the whole trimmer as a compounded rigid body. If a member of an aggregate can freely rotate, then the hydrodynamic stresses exerted on the outer surface of the compound body must be relaxed, allowing rotation of individual spheres, and therefore the net shear stress is reduced. If the same shear force is applied to the rigid trimmer, then the non-holonomic constraint strongly resists the external hydrodynamic stress and adjusts its position to minimize the external stress. The angles made by connecting three particle centers in the last snapshots in **Figure 5(a)** and **(b)** indicate the different responses of the settling trimmer to the hydrodynamic drags and stresses. In addition to these hydrodynamic forces and torques, inelastic properties of granules significantly influence their transient rotating patterns.

3.3. Parallel algorithms

Granular dynamics simulations in a fluid medium must be an open problem in state-of-the-art computational research. Since the length scale of the forces acting on touching granules

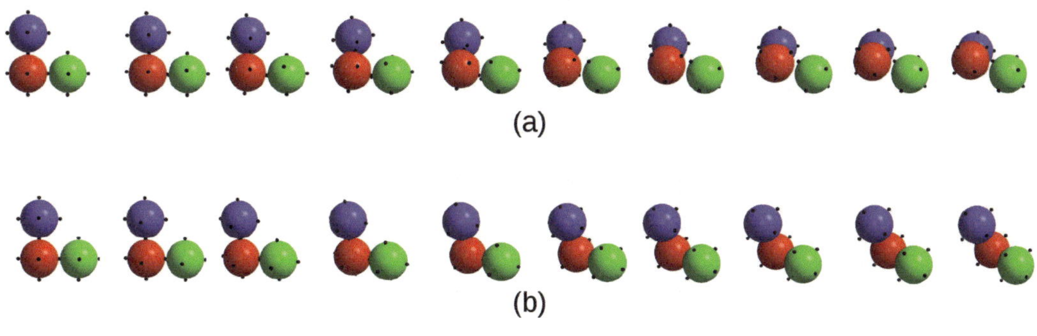

Figure 5. A spherical trimer of (a) holonomic constraint only and (b) holonomic and non-holonomic constraints. In each case, the center-to-center distances between three pairs of particles are fixed.

is much smaller than the granular sizes, simulations seem to be very efficient if parallel algorithms are adequately used. Domain decomposition scheme is one of the most widely used parallel algorithms. The system is divided into several small sub-domains, and a dynamic simulation in a spatial sub-domain is conducted by an individual computing unit, such as a core. This decomposition method is very efficient if particles are evenly distributed in space, which is usual in equilibrium simulations of MD and Monte Carlo. As granules are highly subjected to the gravitational force and hydrodynamic interactions, spatial biases are almost inevitable in both their transient and stationary states. Granular dynamics has at least three length scales of different orders of magnitude. The mutual compression distance, i.e., inter-particle overlap distance, is at least three or four orders of magnitude smaller than the particle size. In a fluid medium, hydrodynamic interactions are long-ranged and quite significant when the surface-to-surface distance between nearest neighbors is about the particle diameter. Motion of heavy granules in a fluid flow may distort the ambient flow-field as well as the hydrodynamic forces exerted on adjacent particles. In granular dynamics, when the contact force exerted between two colliding particles, the force acting on particle i from j has the same magnitude and opposite direction to that from particle j from particle i. Therefore, the number of contact force calculations can be reduced to a half, if Newton's third law is implemented during parallel computing. In this case, the computational efficiency of granular dynamics simulation can be as much as doubled.

4. Concluding remarks

In nature, phases of matter have been conventionally believed to be those of gas, liquid, and solid, in which specific phase transformations are possible between the states. A recent addition of the plasma state has increased the total number of material states from three to four. It is questionable to predict that the granular state will be the fifth matter phase.

On the other hand, the granules dynamically change the representative phase based on external influences. Stationary and compressed granules behave similar to amorphous solids or gel materials. Moving like a fluid, mud or sediments create their own pathways by minimizing hydrodynamic influences. A fast flow with granular materials, such as in streams and ocean, creates a dense turbidity flow, but a decelerating flow field initiates a granular phase change from a flowing liquid to a packed solid. Small dry granules behave similar to dust in the wind, for which standard gas transport theory can predict dynamic behaviors of granular gases. In our opinion, granules are chameleon materials, transitioning their phases dynamically. No equilibrium exists in a granular phase so that thermodynamic fluctuation of the granular state cannot provide material or phase constants. As granular dynamics in transient force/torque fields significantly removes steady-state behaviors, the initial and boundary conditions become extraordinarily important in analyzing many-body granular motion. To utilize specific behavior of granules, granules can be dynamically controlled in vibrating, oscillating, or swinging phases. The time-correlation scale of dissipating granules is not short enough to use the Markovian chain concept because granular paths in both the real and phase spaces significantly influence their fates.

Acknowledgements

This work was financially supported by the National R&D project of "Development of 1MW Ocean Thermal Energy Conversion Plant for Demonstration" (PMS3680) from the Korean Ministry of Oceans and Fisheries and used the Extreme Science and Engineering Discovery Environment (XSEDE), which is supported by National Science Foundation grant number ACI-1053575.

Author details

Albert S. Kim[1]* and Hyeon-Ju Kim[2]

*Address all correspondence to: albertsk@hawaii.edu

1 Civil and Environmental Engineering, University of Hawaii at Manoa, Honolulu, HI, United States of America

2 Seawater Utilization Plant Research Center, Korea Research Institute of Ships and Ocean Engineering, Goseong-gun, Gangwon-do, Republic of Korea

References

[1] Landau LD, Lifshitz EM. Mechanics. New York: Pergamon Press; 1976

[2] Kim S, Karrila SJ. Microhydrodynamics, Principles and Selected Applications. Mineola, New York: Dover Publication, Inc.; 1991

[3] Allen MP, Tildesley DJ. Computer Simulation of Liquids. New York, NY: Clarendon Press; 1987

[4] Jones JE. On the Determination of molecular fields. II. From the equation of state of a gas. Proceedings of the Royal Society A: Mathematical, Physical and Engineering Sciences. 1924;**106**(738):463–477

[5] Verlet L. Computer "experiments" on classical fluids. I. Thermodynamical properties of Lennard-Jones molecules. Physical Review. 1967;**159**(1):98–103

[6] Hockney RW. The Potential Calculation and Some Applications. Methods in Computational Physics. 9. New York: Academic Press; 1970

[7] Swope WC, Andersen HC, Berens PH, Wilson KR. A computer simulation method for the calculation of equilibrium constants for the formation of physical clusters of molecules: Application to small water clusters. The Journal of Chemical Physics. 1982;**76**(1):637–649

[8] Cornell WD, Cieplak P, Bayly CI, Gould IR, Merz KM, Ferguson DM, et al. A second generation force field for the simulation of proteins, nucleic acids, and organic molecules. Journal of the American Chemical Society. 1995;**117**(19):5179–5197

[9] Brooks BR, Bruccoleri RE, Olafson BD, States DJ, Swaminathan S, Karplus M. CHARMM: A program for macromolecular energy, minimization, and dynamics calculations. Journal of Computational Chemistry. 1983;**4**(2):187–217

[10] Van Der Spoel D, Lindahl E, Hess B, Groenhof G, Mark AE, Berendsen HJC. GROMACS: Fast, flexible, and free. Journal of Computational Chemistry. 2005;**26**(16):1701–1718

[11] Phillips JC, Braun R, Wang W, Gumbart J, Tajkhorshid E, Villa E, et al. Scalable molecular dynamics with NAMD. Journal of Computational Chemistry. 2005;**26**(16):1781–1802

[12] Brown R. A Brief Account of Microscopical Observations made in the Months of June, July, and August, 1827, on the Particles contained in the Pollen of Plants; and on the General Existence of Active Molecules in Organic and Inorganic Bodies

[13] Einstein A. Zur Theorie der Brownschen Bewegung. Annalen der Physik. 1906;**324**(2):371–381

[14] Langevin P. Sur la théorie du mouvement brownien. Comptes-rendus de I'Académie des Sciences (Paris). 1908;**146**:530–533

[15] Chandrasekhar S. Stochastic problems in physics and astronomy. Reviews of Modern Physics. 1943;**15**(1):1–89

[16] Chandrasekhar S. Brownian motion, dynamical friction, and stellar dynamics. Reviews of Modern Physics. 1949;**21**(3):383–388

[17] Kubo R. The fluctuation-dissipation theorem. Reports on Progress in Physics. 1966;**29**: 255–284

[18] Stokes GG. On the effect of internal friction of fluids on the motion of pendulums. Transactions of the Cambridge Philosophical Society. 1851;**9**:1–106

[19] Ermakova L, Sidorova M, Savina I, Aleksandrov D. Structural and electrochemical parameters of asymmetric membranes for reverse osmosis. Colloids and Surfaces A. 1998;**142**(2–3):265–274

[20] Ermak DL, Buckholz H. Numerical integration of the Langevin equation: Monte Carlo simulation. Journal of Computational Physics. 1980;**35**:169–182

[21] Ermak DL, Mccammon JA. Brownian dynamics with hydrodynamic interactions. Journal of Chemical Physics. 1978;**69**(4):1352–1360

[22] Ermak D. A computer simulation of charged particles in solution. I. Technique and equilibrium properties. Journal of Chemical Physics. 1975;**62**(10):4189–4196

[23] Ermak DL. A computer simulation of charged particles in solution. I. Technique and equilibrium properties. Journal of Chemical Physics. 1975;**62**(10):4189–4196

[24] Chen JC, Elimelech M, Kim AS. Monte Carlo simulation of colloidal membrane filtration: Model development with application to characterization of colloid phase transition. Journal of Membrane Science. 2005;**255**(1):291–305

[25] Hoogerbrugge P, Koelman J. Simulating microscopic hydrodynamic phenomena with dissipative particle dynamics. Europhysics Letters. 1992;**19**:155

[26] Koelman J, Hoogerbrugge P. Dynamic simulations of hard-sphere suspensions under steady shear. Europhysics Letters. 1993;**21**:363

[27] Klingenberg DJ. Simulation of the dynamic oscillatory response of electrorheological suspensions: Demonstration of a relaxation mechanism. Journal of Rheology. 1993;**37**(2): 199–214

[28] Espanol P, Warren P. Statistical mechanics of dissipative particle dynamics. Europhysics Letters. 1995;**30**(4):191–196

[29] Groot RD, Warren PB. Dissipative particle dynamics: Bridging the gap between atomistic and mesoscopic simulation. Journal of Chemical Physics. 1997;**107**(11):4423–4435

[30] Martys NS, Mountain RD. Velocity Verlet algorithm for dissipative-particle-dynamics-based models of suspensions. Physical Review E. 1999;**59**(3):3733LP-3736 LP

[31] Nikunen P, Karttunen M, Vattulainen I. How would you integrate the equations of motion in dissipative particle dynamics simulations? Computer Physics Communications. 2003;**153**(3):407–423

[32] Satoh A, Chantrell RW, Coverdale GN. Brownian dynamics simulations of ferromagnetic colloidal dispersions in a simple shear flow. Journal of Colloid and Interface Science. 1999;**209**:44–59

[33] Ramezani M, Shamsara J. Application of DPD in the design of polymeric nano-micelles as drug carriers. Journal of Molecular Graphics and Modelling. 2016;**66**:1–8

[34] Karatrantos A, Clarke N, Composto RJ, Winey KI. Topological entanglement length in polymer melts and nanocomposites by a DPD polymer model. Soft Matter. 2013;**9** (14):3877

[35] Prhashanna A, Khan SA, Chen SB. Micelle morphology and chain conformation of triblock copolymers under shear: LA-DPD study. Colloids and Surfaces A: Physicochemical and Engineering Aspects. 2016;**506**:457–466

[36] Kim S, Karrila SJ. Microhydrodynamics: Principles and Selected Applications. New York: Dover Publications; 2005

[37] Bossis G, Brady JF. Dynamic simulation of sheared suspensions. I. General method. Journal of Chemical Physics. 1984;**80**(10):5142–5154

[38] Bossis G, Brady JF. Self-diffusion of Brownian particles in concentrated suspensions under shear. Journal of Chemical Physics. 1987;**87**:5437

[39] Durlofsky L, Brady JF, Bossis G. Dynamic simulation of hydrodynamically interacting particles. Journal of Fluid Mechanics. 1987;**180**:21–49

[40] Brady JF, Bossis G. Stokesian dynamics. Annual Review of Fluid Mechanics. 1988;**20**:111–157

[41] Brady JF, Phillips RJ, Lester JC, Bossis G. Dynamic simulation of hydrodynamically interacting suspensions. Journal of Fluid Mechanics. 1988;**195**:257–280

[42] Wang M, Brady JF. Spectral ewald acceleration of stokesian dynamics for polydisperse suspensions. Journal of Computational Physics. 2016;**306**:443–477

[43] Sierou A, Brady JF. Accelerated stokesian dynamics simulations. Journal of Fluid Mechanics. 2001;**448**

[44] Kim AS. Dissipative hydrodynamics of rigid spherical particles. Chemistry Letters. 2012;**41**(10):1128–1130

[45] Wiener N. Differential Space. Journal of Mathematical Physics. 1923;**58**:31–174

[46] Ito M. An extension of nonlinear evolution equations of the K-dV (mK-dV) type to higher orders. Journal of the Physical Society of Japan. 1980;**49**:771–778

[47] Kim AS, Stolzenbach KD. The permeability of synthetic fractal aggregates with realistic three-dimensional structure. Journal of Colloid and Interface Science. 2002;**253**(2):315–328

[48] Kim AS, Stolzenbach KD. Aggregate formation and collision efficiency in differential settling. Journal of Colloid and Interface Science. 2004;**271**(1):110–119

[49] Kim AS, Kang ST. Microhydrodynamics simulation of Single-collector granular filtration. Chemistry Letters. 2012;**41**(10):1288–1290

[50] Hoomans BPB. Granular Dynamics of Gas-Solid Two-Phase Flows. Universiteit Twente; Enschede, Netherlands; 2000

[51] Poschel T, Schwager T. Computational Granular Dynamics, Models and Algorithms. Springer, Berlin; 2005

[52] Verwey EJ, Overbeek JTG. Theory of the Stability of Lyophobic Colloids. Amsterdam: Elsevier; 1948

[53] Derjaguin BV, Landau L. Theory of the stability of strongly charged lyophobic sols and of the adhesion of strongly charged particles in solutions of electrolytes. Acta Physicochim USSR. 1941;**14**:633–662

[54] Kim AS. Constraint dissipative hydrodynamics (HydroRattle) algorithm for aggregate dynamics. Chemistry Letters. 2012;**41**(10):1285–1287

[55] Hertz H. Ueber die Verdunstung der Flussigkeiten, insbesondere des Quecksilbers, im luftleeren Raume. Annalen der Physik. 1882;**253**(10):177–193

[56] Brilliantov NV, Spahn F, Hertzsch JM, Poschel T. Model for collisions in granular gases. Physical Review E. 1996;**53**(5):5382–5392

[57] Brilliantov NV, Spahn F, Hertzsch JM, Poschel T. The collision of particles in granular systems. Physica A: Statistical Mechanics and its Applications. 1996;**231**(4):417–424

[58] Ramiraz R, Poschel T, Brilliantov NV, Schwager T. Coefficient of restitution of colliding viscoelastic spheres. Physical Review E. 1999;**60**(4):4465–4472

[59] Rosato A, Strandburg KJ, Prinz F, Swendsen RH. Why the Brazil nuts are on top: Size segregation of particulate matter by shaking. Physical Review Letters. 1987;**58**(10):1038

[60] Hong DC, Quinn PV, Luding S. Reverse Brazil nut problem: Competition between percolation and condensation. Physical Review Letters. 2001;86(15):3423

[61] Walliser H. Comment on "Reverse Brazil Nut Problem: Competition between Percolation and Condensation". arXiv preprint cond-mat/0111559. 2001

[62] Mohamed Khayetbius ME, Lauderdale BE, Nagel SR, Jaeger HM. Brazil-nut effect: Size separation of granular particles. Nature. 2001;**414**(6861):270–270

[63] Canul-Chay G, Belmont P, Nahmad-Molinari Y, Ruiz-Suárez J. Does the reverse Brazil nut problem exist? Physical Review Letters. 2002;**89**(18):189601

[64] Breu AP, Ensner HM, Kruelle CA, Rehberg I. Reversing the Brazil-nut effect: Competition between percolation and condensation. Physical Review Letters. 2003;**90**(1):014302

[65] Geisel TS. Bartholomew and the Oobleck. Random House Books for Young Readers. New York; 1949

[66] Roché M, Myftiu E, Johnston MC, Kim P, Stone HA. Dynamic fracture of nonglassy suspensions. Physical Review Letters. 2013;**110**(14)

[67] Tian T, Nakano M. Design and testing of a rotational brake with shear thickening fluids. Smart Materials and Structures. 2017;**26**(3):035038

[68] Wagner NJ, Brady JF. Shear thickening in colloidal dispersions. Physics Today. 2009;**62**(10):27–32

[69] Seto R, Mari R, Morris JF, Denn MM. Discontinuous shear thickening of frictional Hard-Sphere suspensions. Physical Review Letters. 2013;**111**(21).

[70] Hoffman RL. Discontinuous and dilatant viscosity behavior in concentrated suspensions. II. Theory and experimental tests. Journal of Colloid and Interface Science. 1974;**46**(3): 491–506

[71] Brown E, Jaeger HM. The role of dilation and confining stresses in shear thickening of dense suspensions. Journal of Rheology. 2012;**56**(4):875–923

[72] Cui M, Emrick T, Russell TP. Stabilizing liquid drops in nonequilibrium shapes by the interfacial jamming of nanoparticles. Science. 2013;**342**(6157):460–463

[73] Brown E, Jaeger HM. Shear thickening in concentrated suspensions: Phenomenology, mechanisms and relations to jamming. Reports on Progress in Physics. 2014;**77**(4):046602

[74] Humphrey W, Dalke A, Schulten K. VMD: Visual molecular dynamics. Journal of Molecular Graphics. 1996;**14**(1):33–38

[75] Sharma R, Zeller M, Pavlovic VI, Huang TS, Lo Z, Chu S, Zhao Y, Phillips JC, Schulten K. Speech/Gesture interface to a Visual-Computing environment. IEEE Computer Graphics and Applications. 2000;**20**:29–37

[76] Eargle J, Wright D, Luthey-Schulten Z. Multiple Alignment of protein structures and sequences for VMD. Bioinformatics. 2006;**22**(4):504–506

[77] Stukowski A. Visualization and analysis of atomistic simulation data with OVITO–the Open visualization tool. Modelling and Simulation in Materials Science and Engineering. 2010;**18**(1):015012

Chapter 3

Granular Flow: From Dilute to Jammed States

Hao Shi, Dalila Vescovi, Abhinendra Singh,
Sudeshna Roy, Vanessa Magnanimo and
Stefan Luding

Additional information is available at the end of the chapter

http://dx.doi.org/10.5772/intechopen.68465

Abstract

Particulate systems and granular matter display dynamic or static, fluid- or solid-like states, respectively, or both at the same time. The mystery of bridging the gap between the particulate, microscopic state and the macroscopic, continuum description is one of the challenges of modern research. This book chapter gives an overview of recent progress and some new insights about the collective mechanical behavior of granular, deformable particles.

Keywords: rheology, solid-fluid granular behavior, micro-macro transition, numerical simulations

1. Introduction

Dune migration, landslides, avalanches, and silo instability are a few examples of systems where granular materials play an important role. Furthermore, handling and transport of these materials are central to many industries such as pharmaceutical, agricultural, mining, and construction and pose many open questions to the researchers. In spite of their ubiquity, understanding and predicting the flow behavior of granular materials is still a major challenge for science and industry. Even in a seemingly simple system such as dry sand, the presence of large numbers of internal degrees of freedom leads to highly nonlinear effects making it difficult to relate the microscopic grain-level properties to the macroscopic bulk behavior.

Granular systems can show properties commonly associated with either solid or liquid. They can behave like a fluid, that is, yielding under an applied shear stress. On the other hand, they can also behave like solids, being able to resist applied stresses without deforming, showing also interesting anisotropic structure (contact-and force-networks) [1, 2]. Lucretius (ca. 98–55 B.C.) was among the first ones to recognize this interesting behavior of soil-like materials, when he wrote "*One can scoop up poppy seeds with a ladle as easily as if they were water and, when dipping the ladle, the seeds flow in a continuous stream*" [3].Granular materials exhibit solid-like behavior if the particles are packed densely enough and a network of persistent contacts develops within the medium, resulting in a mechanically stable jammed structure of the particles. On the other hand, when the grains are widely spaced and free to move in any direction, interacting only through collisions, the medium is unjammed and behaves like a fluid [4].

Due to their microscopic, discrete nature and their interesting macroscopic, bulk behavior response, granular materials are studied using both discrete and continuum mechanics frameworks. In the realm of the discrete approach, several numerical techniques that are able to reproduce the single particle motions with the given micromechanical properties of the grains have been developed. In such an approach, the dynamic behavior is studied by integrating the Newton's equations of motion for each grain using micromechanical properties and specific interaction law. Following the pioneer work by Goldhirsch [5, 6], several numerical techniques have been developed to obtain continuum fields from discrete particle data.

Using these numerical methods, one can study the flow behavior of the idealized grains, characterized by some specific micromechanical properties, which might not exist in the nature, but is helpful in understanding the underlying physics of their global behavior. In spite of their versatile applicability and benefits, these numerical methods have limitations such as excessive computational requirements, round off or truncation errors, and an intrinsic dynamic that is sometimes not reflecting the experimental reality. On the other hand, continuum models give a macroscopic view to investigate granular material behaviors. Continuum mechanics theories solve the conservation equations for the whole medium, that is, the balance of mass, momentum, and when necessary, energy. Although the balance laws are easily deducible, defining the constitutive relations poses the bigger challenge. The latter relate stresses and strains taking into account the physics of the grain-grain interaction.

The goal of the present book chapter is to study the constitutive behavior of granular systems using particle, numerical simulations, and micro-macro transition. In particular, we focus on the different mechanical responses of a granular material in dense and dilute conditions, corresponding to the fluid and solid behaviors, respectively. In order to systematically analyze the influence of some crucial material parameters, which affect the flow behavior, we focus on an idealized material composed of frictionless, spherical particles, in the absence of any interstitial fluids. Moreover, in order to concentrate on the rheology of particulate systems, disregarding boundary effects, we have considered two system setups which allow simulating steady and homogeneous flows.

This chapter is organized as follows. Section 2 introduces the general rheological framework to describe the flow behavior of granular materials. In the same section, we also briefly review some existing granular rheological models. The particle simulations along with micro-macro transition are introduced in Section 3, where different system setups that are used to study the steady and homogeneous granular flows are shown. Finally, in Section 4, we present a comprehensive comparison of the existing simulation data with frictionless particles in dilute and dense regimes. In the same section, we highlight the effect of various micromechanical properties (coefficient of restitution, polydispersity, and particle stiffness) on the macroscopic fields (stresses and volume fraction). We present a comparison of these results with the theoretical models in two regimes: the kinetic theory in the dilute regime, and a recently proposed generalized rheological model in the dense regime.

2. Granular rheology

2.1. A micromechanical based continuum approach

Despite the fact that granular materials are discontinuous media, their behavior is commonly described by a continuum approach. Continuum mechanics theories solve the conservation equations of the whole medium, that is, the balance of mass, momentum, and when necessary, energy. Although the balance laws are easily deducible, the big challenge is the definition of the constitutive relations, that is, the rheology. The latter captures the macroscopic behavior of the system, incorporating the microscale grain-grain interaction dynamics.

A granular flow can undergo different behaviors depending on both properties at the particle level and the macroscopic characteristic of the flow (i.e., velocity and concentration). At the microscopic level, each particle is characterized by its shape, dimension, material, and contact properties. For the sake of simplicity, in this chapter an assembly of identical spheres, of diameter d, density ρ_p, and equivalent linear contact stiffness k_n is considered. The density of the continuum medium can be computed as the product of the particle density and the volume fraction, ν, defined as the fractional, local volume occupied by the spheres: $\rho = \rho_p \nu$. Given that each grain i moves with velocity $\boldsymbol{v}_i$, the macroscopic velocity of N-particles flow in a volume V can be defined as the average $\boldsymbol{u} = \frac{1}{V}\sum_{i=1}^{N} \boldsymbol{v}_i$. Similarly, we can introduce the strain-rate tensor, calculated as the symmetric part of the velocity gradient. Its off-diagonal components describe the shear rate between two Cartesian directions and are often used as control parameters to describe flow problems. In particular, considering a granular system with mean flow in the x-direction only and sheared along the y-direction, we introduce the shear rate as $\dot{\gamma} = 2\dot{\varepsilon}_{xy} = \partial u_x/\partial y$. Finally, in continuum mechanics, the stress tensor, $\boldsymbol{\sigma}$, represents the manner in which force is internally transmitted. Each component of the stress tensor, σ_{ij} represents the force in the i-direction on a surface with inward pointing normal unit vector in the j-direction. The isotropic part of the stress tensor is the hydrostatic stress or pressure p, while the shear stress τ is proportional to the second invariant of the stress tensor. A detailed description of how to calculate strain rate and stress tensors in the case of granular assemblies will be provided in Section 2.2.

In the framework of continuum mechanics, dimensionless numbers are often introduced in order to describe the material behavior. These dimensionless numbers are defined as the ratio of different time scales or forces, thus signifying the relative dominance of one phenomenon over another.

In the case of granular flows, the macroscopic time scale associated with the shear rate parallel to the flow plays an important role. Then, it is convenient to scale all the quantities using the particle diameter, particle density, and shear rate $\dot{\gamma}$, so that the dimensionless pressure and stiffness are given as $p/\left(\rho_p d^2 \dot{\gamma}^2\right)$ and $k_n/\left(\rho_p d^3 \dot{\gamma}^2\right)$, respectively. On the other hand, when particle deformability becomes relevant, quantities are usually made dimensionless using the particle stiffness; pressure and shear rate are then expressed as $p\ d/k_n$ and $\dot{\gamma}\sqrt{\left(\rho_p d^3/k_n\right)}$. In the following sections, we will see how these dimensionless numbers are used to characterize granular flows in their different regimes, namely fluid-like and solid-like.

2.2. Continuum models

In the early modeling attempts, granular flow is envisaged as existing in either dense solid-like or loose gas-like regimes. Early works using shear cell experiments observed these regimes by varying the shear rate and allowing the bed to dilate or compact. Granular materials exhibit solid-like behavior if the particles are packed densely enough and a network of persistent contacts develops within the medium, resulting in a jammed mechanically stable structure of the particles. On the other hand, when the grains are widely spaced and free to move in any direction, interacting only through collisions, the medium is unjammed and behaves like a fluid [7].

In the fluid-like limit, the system is very dilute and the grains interact mainly through binary, instantaneous, uncorrelated collisions. One of the first rheological models for granular flows in this regime was proposed in 1954 by Bagnold [8]. This empirical model, derived from experiments in two-dimensional plane shear flows, basically states that the stresses are proportional to the square of the strain rate. This simple law, now known as "Bagnold scaling," has been the first to understand the physics of granular dynamics at large deformations and has been verified for dry grains in a number of experimental and numerical studies [9–12]. In the fluid-like regime, the generalization of kinetic theory of granular gases provides a meaningful hydrodynamic description.

On the other hand, when the system is very dense, its response is governed by the enduring contacts among grains, which are involved in force chains; the deformations are extremely slow because the entire network of contacts has to be continuously rearranged (jammed structure). In these conditions, the granular material behaves like a solid, showing an elastic response in which stresses are rate independent. The corresponding flow regime is usually referred to as quasi-static. Slowly deforming quasi-static dense granular material has been mainly investigated in the framework of geo-mechanics. There, the majority of the constitutive models are based on the theories of elasto-plasticity and visco-plasticity [13–16], and many of them have been conceived by starting from the well-known critical state theory [17, 18].

In the transition phase, the grains interact via both force chains and collisions. None of the models cited above is able to deal with this phase-transition of granular materials from a solid-like to a fluid-like state and vice-versa. Intensive studies of the granular rheology at the phase transition have been conducted in the last decades, for example, by Campbell [19], Ji and

Shen [20, 21], and Chialvo et al. [22] using 3D simulations of soft frictional spheres at imposed volume fractions. In these works, the authors derived a flow-map of the various flow regimes and analyzed the transition areas. In particular, they found that, for a collection of particles, the solid-fluid transition occurs in the limit of zero confining pressure at the critical volume fraction ν_c. Then the solid-like regime, in which stresses are independent of shear rate, occurs for volume fractions $\nu > \nu_c$, whereas, at volume fractions $\nu < \nu_c$ the system shows a fluid-like behavior with stresses scaling with the square of the shear rate. In the proximity of the critical volume fraction, a continuous transition between the two extreme regimes takes place, for which the rheological behavior is still not fully understood.

More recently, new theories have been developed to model the phase transition. The French research group GDR-MiDi [23] has suggested that dense granular materials obey a local, phenomenological rheology, known as μ(I)-rheology, that can be expressed in terms of relations between three nondimensional quantities: volume fraction, shear to normal stress ratio, usually called μ, and inertial parameter I. The latter is defined as the ratio of the time scales associated with the motion perpendicular and parallel to the flow: $I = \dot{\gamma} d \sqrt{\rho_p / p}$ [24, 25]. The inertial number provides an estimate of the local rapidity of the flow, with respect to pressure, and is of significance in dynamic/inertial flows, as shown in Ref. [26]. In dense, quasi-static flows, particles interact by enduring contacts and inertial effects are negligible, that is I goes to zero. Two main assumptions on the basis of the μ(I)-rheology are: (i) perfectly rigid (i.e., nondeformable) particles and (ii) homogeneous flow. Various constitutive relations, based on the GDR-MiDi rheology, have been developed [9, 27–29] in order to extend the validity of the model. In particular, the influence of particle deformability has been accounted for in the soft μ(I)-rheology proposed in Refs. [30–32].

Below we present a summary of the two continuum theories that well describe the flow behavior in the limits and their extension to the intermediate regime. Kinetic theory in its standard form (SKT) provides a meaningful hydrodynamic description for frictionless particles in the very dilute regime, while μ(I)-rheology holds for both frictionless and frictional particles for dense flows. It is important to mention that both theories work only for ideal systems, made of rigid, perfectly elastic, monodisperse particles. Finally, the extension of μ(I)-rheology to deal with soft and deformable particles is also introduced.

2.2.1. *Standard kinetic theory (SKT)*

This section is largely based on the notable works of Brilliantov et al. [33], Garzo et al. [34, 35], Goldhirsch [6, 36], and Pöschel et al. [37].

The term "granular gas" is used in analogy with a (classical) molecular gas, where the molecules are widely separated and are free to move in all directions, interacting only through instantaneous, uncorrelated collisions. The main differences between molecular and granular gases are that in the latter case part of the energy is irreversibly lost whenever particles interact and the absence of strong scale separation. These facts have numerous consequences on the rheology of granular gases, one of which being the sizeable normal stress differences [38].

Analogous to the molecular gases (or liquids), the macroscopic fields velocity and mass density are defined for granular systems [6]. An additional variable of the system, the granular temperature, T, is introduced as the mean square of the velocity fluctuations of the grains, in analogy with molecular gases, quantitatively describing the degree of agitation of the system.

Following the statistical mechanics approach, the kinetic theory of granular gases rigorously derives the set of partial differential equations given by the conservation laws of mass, momentum, and energy (the latter describing the time development of the granular temperature) for the dilute gas of inelastically colliding particles.

In this section, we summarize the standard kinetic theory (SKT) for the case of steady and homogeneous flows for a collection of ideal particles, that is, they are rigid, monodisperse, frictionless with diameter, d, and density, ρ_p. In this case, the mass balance is automatically satisfied, the momentum balance trivially asserts that the pressure, p, and the shear stress, τ, are homogeneous and the flow is totally governed by the balance of energy, which reduces to

$$\Gamma = \tau\,\dot{\gamma} \tag{1}$$

where Γ is the rate of energy dissipation due to collisions and γ is the shear rate. The constitutive relations for p, τ, and Γ are given as [39]

$$\begin{aligned} p &= \rho_p f_1\, T \\ \tau &= \rho_p\, d f_2 T^{1/2}\dot{\gamma} \\ \Gamma &= \frac{\rho_p}{d} f_3 T^{3/2} \end{aligned} \tag{2}$$

where, f_1, f_2, and f_3, are explicit functions of the volume fraction ν and the coefficient of restitution, e_n, (ratio of precollisional to postcollisional relative velocity between colliding particles in the normal impact direction), and are listed in **Table 1**.

$$f_1 = 4\nu G F$$

$$f_2 = \frac{8J}{5\pi^{1/2}}\nu G$$

$$f_3 = \frac{12}{\pi^{1/2}}\left(1 - e_n^2\right)\nu G$$

$$G = \nu\frac{(2-\nu)}{2(1-\nu)^3}$$

$$F = \frac{(1+e_n)}{2} + \frac{1}{4G}$$

$$J = \frac{(1-e_n)}{2} + \frac{\pi}{32}\frac{[5+2(1+e_n)(3e_n-1)G][5+4(1+e_n)G]}{\left[24-6(1+e_n)^2-5\left(1-e_n^2\right)\right]G^2}$$

Table 1. List of coefficients as introduced in the constitutive relations of SKT (standard kinetic theory).

Further, by substituting the constitutive relations for τ and Γ into the energy balance, the granular temperature drops out, so that the pressure becomes proportional to the square of the strain rate (Bagnold scaling [8])

$$p = \rho_p d^2 f_1 \dot{\gamma}^2 \tag{3}$$

SKT was rigorously derived under very restrictive assumptions. In particular, the granular system is assumed to be monodisperse and composed of spherical, frictionless, and rigid particles, interacting only through binary and uncorrelated collisions [7, 40, 41]. Several modifications to the SKT have been introduced in the literature accounting for different effects: interparticle friction [4, 7, 42–44], nonsphericity [45], or polydispersity [46]. As one example, Jenkins [47, 48] extended the kinetic theory to account for the existence of correlated motion among particles at high concentration.

2.2.2. *Traditional μ(I) rheology*

A convincing, yet simple phenomenological model that predicts the flow behavior in moderate-to-dense regime is the μ(I) rheology. Once again, this rheological law is based on the assumption of homogeneous flow of idealized rigid, monodisperse particles, though the extra constraint of frictionless particles can be dropped. According to this empirical model, only three dimensionless variables are relevant for steady shear flows of granular materials: the volume fraction ν, the shear stress to normal stress ratio $\mu = \tau/p$, and the inertial number I [9, 23, 28]. The collaborative study GDR-Midi showed the data collapse for various shear geometries such as inclined plane, rotating drum, and annular shear when analyzed in terms of the inertial number. μ(I) rheology in the standard form is given by

$$\mu = \mu_0 + \frac{(\mu_\infty - \mu_0)}{I_0/I + 1} \tag{4}$$

with μ_0, μ_∞, and I_0 being dimensionless, material parameters which are affected by the micromechanical properties of the grains [49].

To account for the polydispersity of particles, the generalized inertial number taking into account the average diameters of the particles was introduced by [50]. Traditional μ(I) rheology had been successful in describing the flow behavior of homogeneous flows (both dense and fast). But it has failed to capture the slow and nonhomogeneous flow, where a shear rate gradient is present. Researchers have made significant efforts into developing nonlocal models for granular flows [51].

2.2.3. *Soft μ(I) rheology*

When particles are not perfectly rigid, instead they have a finite stiffness (or softness), the binary collision time is nonzero and hence presents an additional timescale, which is ignored in the standard inertial number phenomenology. A dimensionless number signifying the finite softness of the particles is the dimensionless pressure $p^* = pd/k_n$, which is needed to describe the flow behavior, as proposed recently in Refs. [30–32].

$$\mu(I, p^*) = \mu(I)\left(1 - \left(\frac{p^*}{p_0^*}\right)^{0.5}\right) \tag{5}$$

with the dimensionless pressure p^* being the characteristic pressure at which this correction becomes considerable.

The other dimensionless number needed for the full flow characterization is the volume fraction ν. In case of rigid particles under shear, the packing will dilate and hence ν depends only on the inertial number I. On the other hand, a packing made up of soft particles will dilate due to shear, at the same time pressure will lead the compression of the particles. Hence ν depends on both I and p^* as

$$\nu(I, p^*) = \nu_c\left(1 + \frac{p^*}{p_c^*}\right)\left(1 - \frac{I\sqrt{\nu}}{I_c}\right) \tag{6}$$

where I_c and p_c^* are material dependent dimensionless quantities [49, 52] and ν_c is the critical volume fraction, governing the fluid-solid transitions introduced in the previous section. Its dependence on the polydispersity of the system will be discussed in Section 4.

3. Numerical simulations

Since a few decades, dynamic particle simulations have been a strong tool to tackle many challenging issues related to understanding the flow behavior of particulate systems.

The molecular dynamics or discrete element methods (DEM) is the term given to the numerical procedure, which is used to simulate assemblies of discrete particles. Molecular dynamics (MD) was originally introduced to simulate the motion of molecules [53–55]. It is essentially the simultaneous numerical solution of Newton's equation for the motion of individual particles, for which the position, velocity, and acceleration are computed at each time step. Through averaging of positions, velocities, and forces of the particles, the macroscopic fields of the whole system, such as the density, mean velocity, and stresses can be obtained in terms of the micromechanical properties. This helps in revealing insights of the behavior of granular materials, which cannot be captured by experiments. In particular, with MD methods, the role of micromechanical properties of the grains on the macroscopic collective behavior of the system can be analyzed.

Particle simulation methods include three different techniques: The discrete element method (DEM), the event-driven (ED), and the contact dynamics method (CD). All these methods simulate the inelastic and frictional nature of the contacts among grains through microscopic coefficients (i.e., the coefficients of restitutions and the interparticle friction coefficient). In DEM, deformations of particles during contacts are modeled allowing a finite overlap between grains, whereas in the other two methods, the particles are assumed to be infinitely rigid. Since the results presented in this chapter are obtained by using DEM simulations, below we briefly present an overview of DEM. Readers interested in the latter two methods are referred to Refs. [56–58].

3.1. Discrete element method (DEM)

The discrete element method (DEM) is a family of numerical methods for simulating the motion of large numbers of particles. In DEM, the material is modeled as consisting of finite number of discrete particles, with given micromechanical properties. The interactions between particles are treated as dynamic processes with states of equilibrium developing when the internal forces balance. As previously stated, the granular material is considered as a collection of discrete particles interacting through contact forces. Since the realistic modeling of the deformations of the particles is extremely complicated, the grains are assumed to be nondeformable spheres which are allowed to overlap [58]. The general DEM approach involves three stages: (i) detecting the contacts between elements; (ii) calculating the interaction forces among grains; and (iii) computing the acceleration of each particle by numerical integrating the Newton's equations of motion while combining all interaction forces. This three-stage process is repeated until the entire simulation is complete. Based on the fundamental simulation flow, a large variety of modified codes exist and often differ only in terms of the contact model and some techniques used in the interaction force calculations or the contact detection.

In this chapter, we focus on the standard linear spring-dashpot (LSD) model. Considering two particles, i and j, of diameter d and density ρ_p (i.e., mass $m = \rho_p \pi d^3/6$), their contact leads to the normal (in the direction connecting the centers of the two particles in contact) and tangential components of forces as

$$F_{ij}^n = -k_n \delta_{ij}^n - \eta_n \dot{\delta}_{ij}^n \quad F_{ij}^t = -k_t \delta_{ij}^t - \eta_t \dot{\delta}_{ij}^t \tag{7}$$

where δ_{ij}^n and δ_{ij}^t are the normal and tangential component of the overlap at the contact among particle i and particle j, k_n, and k_t the spring stiffness constants, and η_n and η_t the viscous damping coefficients, representing the energy dissipation at the contact, and dots stand for the time derivative. Tangential force is bounded by the Coulomb criterion $|F_{ij}{}^t| < \mu_p F_{ij}{}^n$ with particle friction coefficient μ_p. The resulting contact force vector is then $\boldsymbol{F_{ij}} = F_{ij}{}^n\, \boldsymbol{n} + F_{ij}{}^t\, \boldsymbol{t}$, being $\boldsymbol{n}$ and $\boldsymbol{t}$ the normal and tangential unit vectors at the contact.

Collisions may be described using the coefficients of normal and tangential restitution, e_n and e_t, respectively, relating the pre-collisional and post-collisional relative velocities. For the spring-dashpot model, the following relations between the coefficients of restitution, the spring constants and the damping coefficients hold [59]

$$\gamma_n = \sqrt{\frac{4mk_n(\log e_n)^2}{\pi^2 + (\log e_n)^2}}, \quad \gamma_t = \sqrt{\frac{8mk_t(\log e_t)^2}{7[\pi^2 + (\log e_t)^2]}}, \quad k_t = \frac{2k_n[\pi^2 + (\log e_t)^2]}{7[\pi^2 + (\log e_n)^2]}. \tag{8}$$

3.2. Micro-macro transition

A research goal in the granular community is to derive macroscopic continuum models based on relevant micromechanical properties. This means to bridge the gap between the microscopic properties and the macroscopic mechanical behavior. The methods and tools for this so-called micro-macro transition are often applied to small so-called representative volume elements (RVEs), where all particles can be assumed to behave similarly. Note that both time- and space-averaging are required to obtain reasonable statistics, the latter being appropriate in the case of steady states.

As previously introduced in Section 2.1, the average velocity of N particles in the RVE V gives the macroscopic velocity $\boldsymbol{u}$, while the strain-rate tensor involves the velocity gradient of the particles

$$\dot{\varepsilon} = \frac{1}{2}\sum_{i=1}^{N}\left(\nabla \boldsymbol{v}_i + \nabla^T \boldsymbol{v}_i\right) \tag{9}$$

being $\boldsymbol{v_i}$ the velocity of particle i. For the particular case of granular systems with mean flow in the x-direction only and subjected to shear in the y-direction, the shear rate is introduced as $\dot{\gamma} = 2\dot{\varepsilon}_{xy}$.

The stress tensor is of particular interest for the description of any continuum medium. In the case of granular assemblies, previous studies have proposed stress-force relationships for idealized granular systems that relate average stress in the assembly to fundamental parameters that are explicitly related to statistical averages of inter-particle load transmission and geometrical arrangement. When referring to a homogeneous volume element V, the macroscopic stress tensor $\boldsymbol{\sigma}$ can be calculated as

$$\boldsymbol{\sigma} = \frac{1}{V}\left[\sum_{i=1}^{N} m\ \mathbf{V}_{\mathrm{i}} \otimes \mathbf{V}_i - \sum_{i=1}^{N}\sum_{j\neq i} \boldsymbol{F}_{ij} \otimes l_{ij}\right], \tag{10}$$

where $\boldsymbol{F}_{ij}$ is the contact force and l_{ij} the branch vector in between connecting the centers of particles i and j, and $\boldsymbol{V}_i = \boldsymbol{v}_i$ - $\boldsymbol{u}$ is the velocity fluctuation of particle i. The first and second terms in the previous equation represent the dynamic and static contributions, respectively [5, 60]. The pressure and shear stress are finally defined as

$$p = \frac{1}{3}(\sigma_1 + \sigma_2 + \sigma_3),\ \tau = \sqrt{\frac{(\sigma_1 - \sigma_2)^2 + (\sigma_1 - \sigma_3)^2 + (\sigma_2 - \sigma_3)^2}{2}} \tag{11}$$

where σ_1, σ_2, σ_3 are the eigenvalues of the stress tensor in Eq. (10). With the development of computational power, nowadays one can simulate reasonable number of particles in a granular system and retrieve good statistical information by micro-macro procedure. The simulations and coarse-graining presented in this section were undertaken using the discrete element method (DEM) open-source code Mercury-DPM (www.mercurydpm.org).

3.3. Simulation setups

There are two popular ways to extract continuum quantities relevant for flow description such as stress, density, and shear rate from the discrete particle data. The traditional one is ensemble averaging of "microscopic" simulations of homogeneous small samples, a set of independent RVEs. A recently developed alternative is to simulate a nonhomogeneous geometry where dynamic, flowing zones and static, high-density zones coexist. By using adequate local averaging over equivalent volume (inside which all particles can be assumed to behave similarly), continuum descriptions in a certain parameter range can be obtained from a single simulation.

In Section 4 we will combine results from (a) simple shear RVE and (b) split-bottom shear cell. The setups are briefly introduced and shown in **Figure 1** (see Refs. [30, 49] for more details) and relevant numerical parameters are reported in **Table 1**. When dimensionless quantities (see Section 2.1) are matched and averaging zones are properly selected, the behaviors from different setups are comparable and a wide flow range can be explored.

3.3.1. Simple shear RVE

The collection of spheres of mean diameter d and density ρ_p, sheared under steady conditions is considered. Here and in the following, x and y are taken to be the flow and the shearing directions, respectively, and variations along the transversal direction z are ignored. We also introduce the polydispersity w as the ratio between the maximum and the minimum particle diameter. In this simple configuration, the flow is assumed to be one-dimensional such that the horizontal velocity u_x is the only nonzero component, and the stress tensor reduces to two scalars; the pressure p and the shear stress τ. In the steady state, the mass balance equation is automatically satisfied and the divergence of the velocity is zero. The momentum balance equation, in absence of external forces, indicates that both pressure p and shear stress τ are constant. Simple shear flows are homogeneous if the horizontal velocity of the medium varies

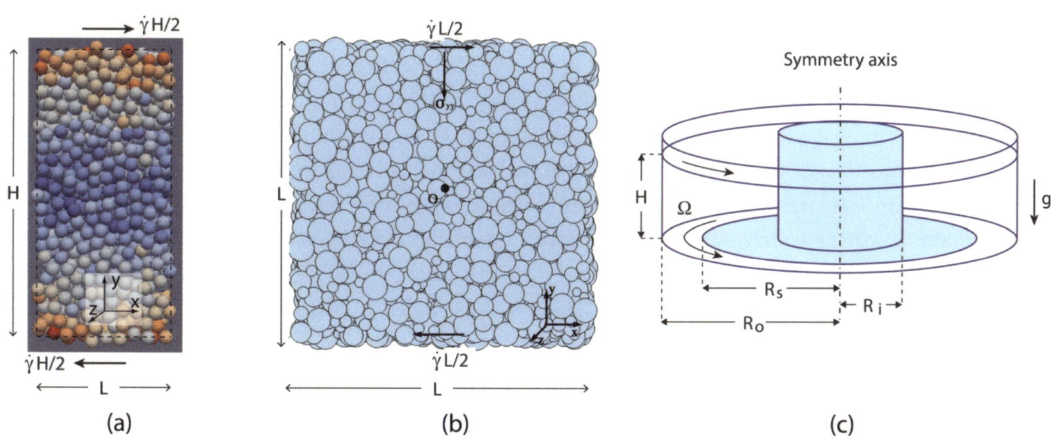

Figure 1. Simulation setups: (a) RVE of monodisperse spheres subjected to constant volume simple shear. The particles have highest kinetic energy near the top and bottom boundaries and lowest near the center in height direction; (b) RVE of polydispersed particles subjected to constant normal stress simple shear; and (c) split-bottom shear cell consisting of a fixed inner part (dark) and a rotating outer part (white).

linearly along the gradient direction and the dominant kinematic variable is its first spatial derivative, the shear rate, $\dot{\gamma} = \partial u_x / \partial y$, which is kept constant along the flow depth. The shear is applied using Lees–Edwards periodic boundary conditions in the y-direction and periodic boundary conditions are employed in the x-and z-directions.

Variables governing the problem are the volume fraction ν (also known as density/concentration defined as the fraction of volume occupied by the spheres), the pressure p, and the shear stress τ. Using DEM simulations, we have performed simulations by using two types of simple shear experiments, that is, (i) constant pressure (here refers to normal stress) or (ii) constant volume boundary conditions. In the former (**Figure 1b**), pressure and strain rate are held constant, hence density and shear stress are outputs and the system is free to dilate/compact based on the initial volume fraction of the packing. In case of constant volume (**Figure 1a**), volume fraction and shear strain rate are held constant, so that pressure and shear stress are the outputs. Constant pressure is one of the traditional methods used in the soil mechanics to estimate the shear strength of the material, while constant volume method is used often to understand the flow behavior close to the jamming transition. Shearing under constant-volume conditions is difficult to realize experimentally due to the fundamental characteristic of the behavior of granular materials, however, a pertinent experiment would be the undrained shear test on water-saturated sand where the volume of the whole specimen can be kept constant within the range of experimental error [18]. On the other hand, dense granular flows under constant stress are present under experimental or natural conditions, for example, sand or/and powders sheared in different shear cells [61] or in an avalanche [62].

Constant-volume steady simple shear samples are placed in a cuboid box (**Figure 1a**). The height of the computational domain as $H = 20d$, with d particle diameter, is fixed before we compute the x- and z-size L according to the chosen, fixed, volume fraction ν. Simulations have been performed using a monodisperse system ($w = 1$) by systematically changing both the volume fraction ν, ranging from dilute to dense regime and the particle stiffness k_n such that the dimensionless shear rate $\gamma(\rho_p d^3/k_n)^{1/2}$ ranges from 3×10^{-2} to 3×10^{-4}.

In the case of RVE under constant normal stress condition (**Figure 1b**), granular systems with polydispersity $w = 2$ and $w = 3$ are considered. The initial length of side is set to L, along with the center point in x-y-plane (marked as O), where one always has zero mean field shear velocity during the whole simulation. The normal stress σ_{yy} is kept constant along y-direction. In this way, the sample is free to dilate/compact along y-direction and smoothly reaches its steady state. In order to investigate the sheared granular flow behavior with different inertia and particle stiffness, we systematically vary both the confined normal stress σ_n and shear strain-rate γ such that the dimensionless stress/softness $\sigma_{yy}(d/k_n)$ ranges between 10^{-3} and 10^{-1} and the dimensionless shear strain-rate $\dot{\gamma}(\rho_p d^3/k_n)^{1/2}$ is between 10^{-5} and 1.

3.3.2. *Split-bottom ring shear cell*

A common feature of natural slow granular flow is the localization of strain in shear bands, which are typically of few particle diameters width. A specialized geometry proposed recently which allows one to impose an external deformation at constant rate is so-called split-bottom geometry (**Figure 1c**). In this geometry, stable shear bands of arbitrary width can be achieved allowing for a detailed study of microstructure associated with the flow of granular materials in the steady state. Unlike the previous setups, in the split-bottom geometry, the granular material is not sheared directly from the walls, but from the bottom. The bottom of the setup that supports the weight of material above it is split in two parts, the two parts move relative to each other and creates a wide shear band away from sidewalls. The resulting shear band is robust, as its location exhibits simple and mostly grain independent properties.

In this geometry, due to inhomogeneous flow, granular packings with contrasting properties and behavior coexist, that is, high-density static to quasi-static areas and dilated dynamic flowing zones are found in the same system. A superimposed grid meshes the granular bed and averaging is performed within each grid volume. Inside a grid volume all particles are assumed to behave similarly and information for a wide parameter range can be obtained using a single numerical experiment, for example, at increasing pressure levels along the depth of the cell. In the following sections, when presenting data from split-bottom cell simulations, only grid-points in the center of the shear band will be considered, where the shear rate γ is higher than a given threshold (see Refs. [3, 30–32, 63] for details on the data processing). Data in center of the shear band are not affected by boundary effects, so that flow gradients can be neglected and the system can be considered as locally homogeneous. In the split-bottom geometry, the shear rate γ is computed as a function of the relative angular velocity Ω between inner and outer cylinders. Details on the geometry setup and numerical parameters adopted for the simulations described in the following section are reported in **Table 2**.

Parameter/Setup	Symbol	Constant Volume	Constant Pressure	Split-Bottom Shear Cell
		Simple Shear	Simple Shear	
Geometry		$L \times H \times L$	$L \times L \times L$	$R_s = 40d$
		$H = 20d$	$L = \text{var}$	$R_i = 0.2R_s$
		$L = \text{var}$		$R_o = 1.3R_s$
				$H = 0.4R_s$
Boundary conditions		Periodic	Periodic	Periodic in azimuthal direction
Number of particles	N	2000	4096	37,000
Polydispersity	w	1	2 and 3	2
Coeff. of restitution	e_n	0.7	0.8	0.8
Volume fraction	ν	0.2–0.68		
Dimensionless stress	$\sigma_{yy}d/k_n$		10^{-3}–10^{-1}	10^{-8}–10^{-2}
Dimensionless shear rate	$\dot{\gamma}(\rho_p d^3/k_n)^{1/2}$	3×10^{-2}–3×10^{-4}	10^{-5}–10^{0}	10^{-2}–10^{-5}

Table 2. Numerical parameters for the three simulation setups.

4. Rheological flow behavior

In this section, we compare the results from various flow setups discussed above for low-to-high volume fractions. We vary various particle and contact properties to understand how the particle micromechanical properties influence the macroscopic flow behavior. We have compared different datasets from different setups and/or authors, and numbered as follows: [A] Peyneau et al. [64]; [B] Chialvo and Sundaresan [65]; [C] Shi et al. (unpublished); [D] Singh et al. [30, 63], and [E] Vescovi and Luding [49]. Unless specified, we will only use the data labels in the following discussion for the sake of brevity.

4.1. Influence of coefficient of restitution

Figure 2 presents a data collection from two different setups and plots the dimensionless pressure against volume fraction. It shows data with constant pressure simulations from data [A] together with the constant volume simulation results of data [B], for frictionless monodisperse rigid

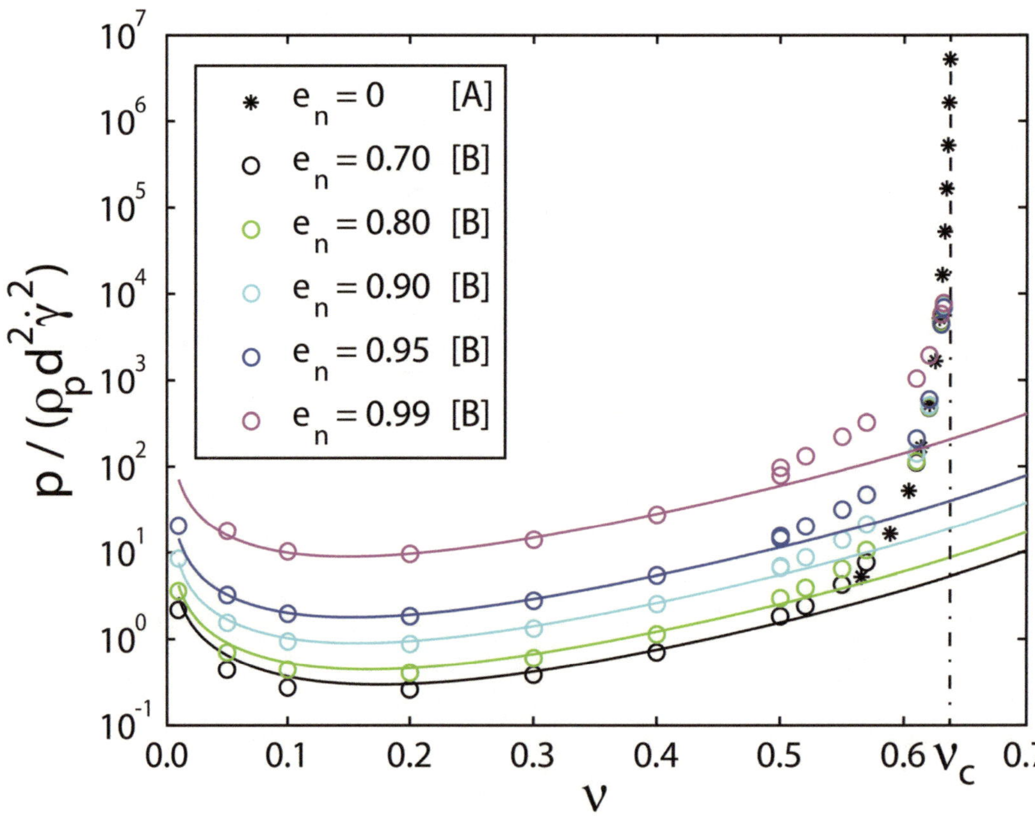

Figure 2. Steady state dimensionless pressure as a function of volume fraction for a simple shear flow of frictionless monodisperse rigid particles. Stars and circles represent simple shear simulations at constant pressure simulations for $e_n = 0$ from data [A] and constant volume fraction for different e_n from data [B], respectively. Different colors refer to different coefficient of restitution as shown in the legend. Different lines are prediction using standard kinetic theory (SKT) as in Eqs. (1–3).

particles. As expected, the data from the two setups are in good agreement. We observe that the restitution coefficient e_n affects the dimensionless pressure strongly for volume fractions $\nu < 0.6$, which increases with increase in e_n. However, in the high volume fraction limit, the data for different e_n collapse on the limit curve diverging at ν_c, that is, ν ranging between 0.6 and the critical volume fraction ν_c.

For the dilute case, a granular gas with high restitution coefficient, for example, $e_n = 0.99$ will behave nearly like an ideal gas, that is, almost no energy loss during each particle-particle collision. Hence, the system will reach equilibrium with higher fluctuation velocity (proportional to the dimensionless pressure) for each particle. In the other extreme, for a restitution coefficient equal to 0, the particles lose all their energy at one collision. Such strong dissipation leads to a rather small pressure in the system. As ν approaches the critical volume fraction, for rigid spheres, the mean free path available for particles decreases making it more difficult to move the particles by imposing shear. The frequency of the collisions and thus the pressure both increase since the free path decreases, diverging in the limit case. Once one reaches the critical volume fraction limit, the system is jammed, hence shear movement of particles without further deformation is not possible. The increase of the pressure for decreasing volume fraction (below 0.1), as the probability of collisions is reduced in the dilute case, is due to the collisional energy loss with a higher steady state pressure. As for the standard kinetic theory prediction, it captures the behavior below volume fractions 0.5 well, but fails for higher volume fractions. This is expected because the standard kinetic theory (SKT) does not take the critical volume fraction into account and thus leads to an underestimation of the pressure for high volume fractions.

4.2. Influence of polydispersity

Figure 3 shows the variation of the nondimensional pressure with volume fraction for different polydispersity for constant pressure (data [A] and [C]), constant volume (data [B]) homogeneous shear flow simulations, together with the local shear band data from nonhomogeneous shear flows (data [D]). We observe that for low-to-moderate volume fractions, pressure is weakly increasing with volume fraction. The data from different shear setups and different polydispersity collapse and agree with the predictions of SKT. However, for higher volume fractions ($\nu > 0.55$), pressure increases when approaching ν_c. However, different polydispersity yields different ν_c [66], so that the pressure decreases with increase in polydispersity, due to the increase in free space available for particle movement for higher polydispersity (in the cases studied here). In some cases, the small particles can move into the gaps between larger particles and form rattlers (rattlers do not contribute to the pressure as for mechanically stable contacts). Therefore, the critical volume fraction ν_c increases with increase in polydispersity as shown by the vertical dashed lines, consistent with previous studies [66–68]. Note that the shear band data from nonhomogeneous split-bottom setup (data [D]) has more scattered than the others, due to the fluctuations of the local averaging over small volumes. But most of the data still follow exactly the same trend as the homogeneous shear data for same polydispersity. We also note that some data points, for example, for polydispersity $w = 3$, go beyond the critical volume fraction due to the fact that DEM particles are not

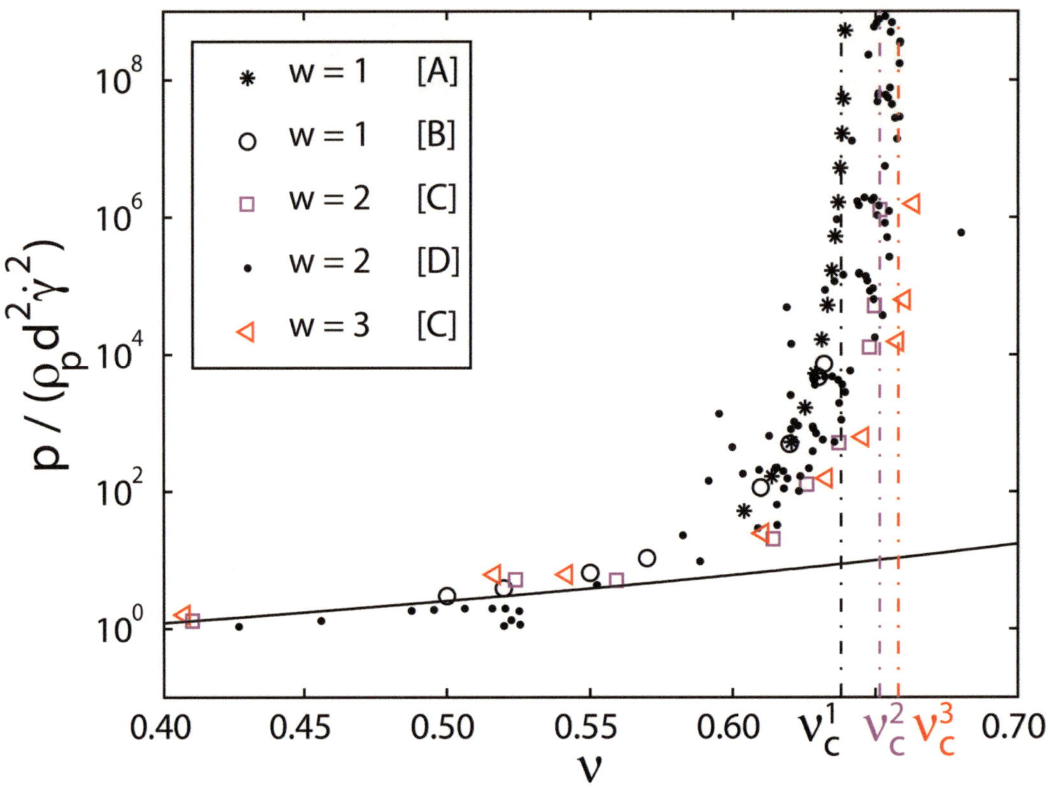

Figure 3. Steady state dimensionless pressure as a function of volume fraction for shear flow simulations of frictionless rigid particles with the same coefficient of restitution ($e_n = 0.8$) but different polydispersity and different setups (data [A–D]) as shown in the legend. The solid line is the prediction of standard kinetic theory and the vertical dashed lines are the predictions of the critical volume fraction with different polydispersity as proposed in Refs. [66–68].

infinitely rigid (they have large but finite stiffness). This softness (and hence possibility of deformation) leads to flow above ν_c and will be elaborated next.

4.3. Effect of particle stiffness

In **Figure 4**, we show the dimensionless pressure as a function of volume fraction for various values of dimensionless particle stiffness, ranging from 10^3 to 10^7. The vertical dashed line shows the monodispersed critical volume fraction as in **Figure 3**. For the sake of comparison, rigid cases (data [A] and [B]) are also plotted. As expected, for the rigid case, pressure diverges close to the critical volume fraction. For soft particles, the deviation from the rigid case is a function of particle stiffness and depending on the system volume fractions (even for the softest particles the deviation from the rigid limit is small for volume fractions smaller than 0.55). When decreasing the volume fraction below 0.5, all different stiffness data tend to collapse. The solid line is the same standard kinetic theory as in **Figure 3** where the assumption of rigid particle breaks down for volume fractions $\nu > 0.5$. And the horizontal dashed line is the prediction from extended rheological model in Eq. (6) using the fitting parameters taken from Ref. [49] for the data with dimensionless particle stiffness 10^5. Our new extended dense

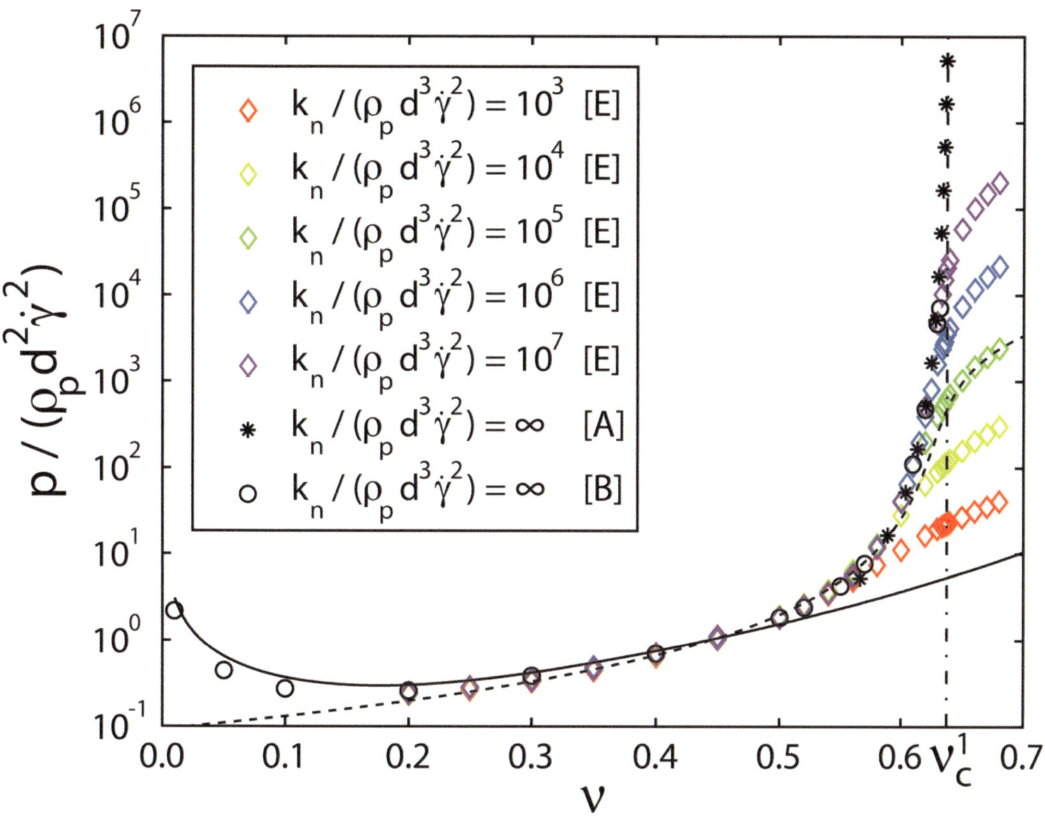

Figure 4. Steady state dimensionless pressure as a function of volume fraction for different values of dimensionless particle stiffness, using monodisperse particles, with restitution coefficient $e_n = 0.7$. Diamonds represent the data from constant volume simulations with data [A], [B], and [E] as suitable for monodisperse, frictionless spheres. The solid line is the prediction of standard kinetic theory and the dashed line is the critical volume fraction as also shown in **Figure 3**. The new horizontal dashed line is the prediction using Eq. (6) with $I_c = 3.28$, $p_c^* = 0.33$, and $\nu_c = 0.636$.

rheological model smoothly captures the soft particles behavior even beyond the critical volume fraction and works perfectly between volume fraction 0.3 and 0.7.

4.4. Combining both particle stiffness and polydispersity in the dense regime

Figure 5 displays dimensionless pressure plotted against volume fraction for both constant volume (data [E]) and normal stress (data [C]) setups with three polydispersities and dimensionless contact stiffnesses, in the moderate to dense volume fraction regime. Diamonds represent constant volume simulation for monodisperse particles while stars and triangles refer to the constant pressure simulation data for polydispersity 2 and 3, respectively, and different color represent different particle stiffness. For $\nu < 0.55$, the data points from the two setups collapse and following the same trend as for the rigid case (**Figure 3**, data [A]). Interestingly, for the data above the critical volume fraction ν_c, the pressure data for different polydispersity are found to collapse with a given dimensionless stiffness (both for 10^5 and 10^7). This indicates that once the system is jammed, the particle stiffness (deformation) determines the

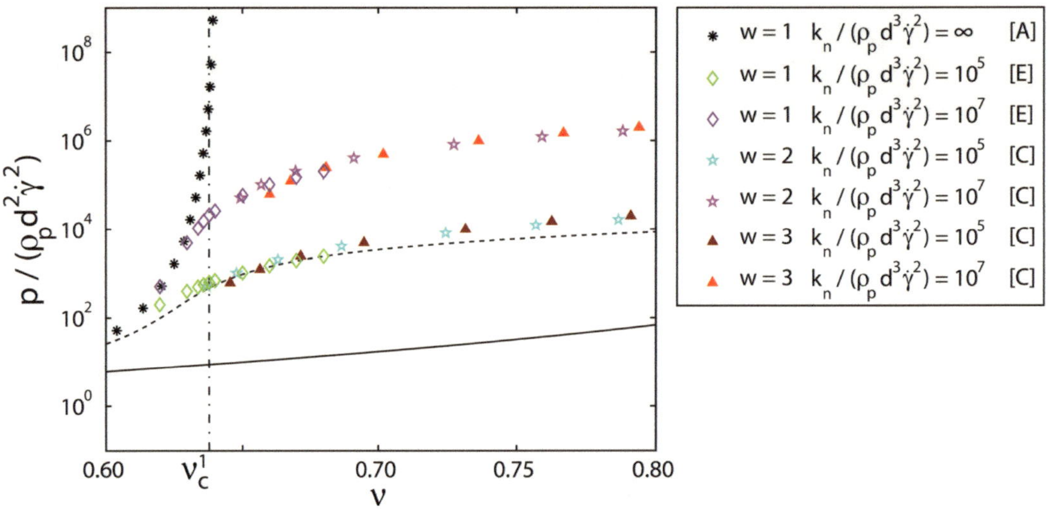

Figure 5. Steady state dimensionless pressure as a function of volume fraction in moderate to dense regime for simulations with different polydispersity and different particle stiffness as given in the legend. The lines are the same as in **Figure 4**.

pressure without much effect of the polydispersity of particles. The solid and dashed lines are the same lines as in **Figure 4**, but is given there as a guidance to the eye representing a reference to the connections. We observe the SKT solid line is not predicting the behavior at all while the extended dense rheology dashed line is qualitatively capturing the behavior even for volume fractions $\nu > 0.7$, but with considerable deviations. Note that there are small differences between the data from two setups and it is due to the small differences in the particles stiffness, and this will be elaborated in the next section.

4.5. From dilute to dense, from "liquid" to "solid," universal scaling

Figure 6 shows the pressure nondimensionalized in two possible ways (a) using shear rate and (b) using particle stiffness (as introduce in Section 2.1) plotted against the distance from the critical volume fraction for the data from different simulations using frictionless particles. **Figure 6a** shows a good data collapse for the volume fractions below the critical volume fraction (unjammed regime), or the so-called fluid regime. In the special case of nearly rigid particles or small confining stress, the scaled pressure diverges at the critical volume fraction, which indicates that the granular fluid composed of rigid particles under shear cannot reach a denser shear jammed state. For the data with softer particles, flow is possible even above the critical volume fraction. For low to moderate volume fractions, the agreement of our data with the rigid case is excellent, while for high volume fractions (especially close to the critical volume fraction) deviations are considerable. The data collapse in the low volume fraction regime shows that the Bagnold scaling relationship between pressure and volume fraction is not strongly affected by particle stiffness, polydispersity, and shear setups, but was influenced by the restitution coefficient (see **Figure 2**). The "fluid" experiences the energy loss more prominent due to collisions.

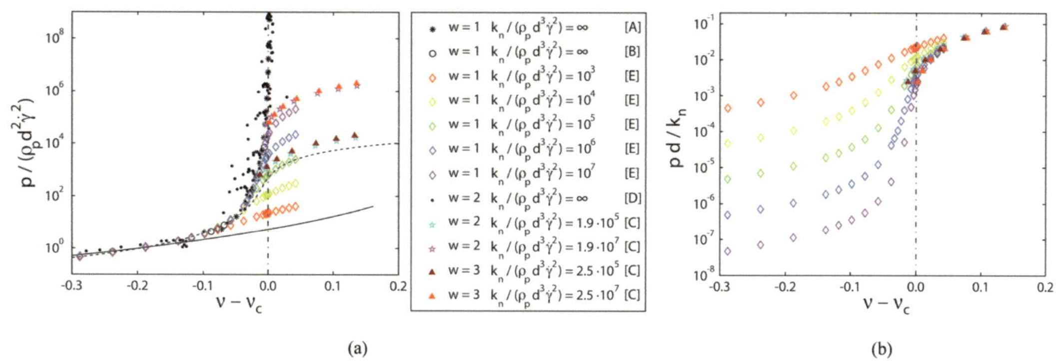

Figure 6. Steady state pressure, as nondimensionalized by (a) shear rate and (b) particle stiffness plotted as a function of distance from the critical volume fraction, ν-ν_c, for frictionless particles in different shear setups with different polydispersity and stiffness as shown in the legend. The lines are the same as in **Figure 4**.

For larger volume fractions, the scaling does not collapse the data. Note the deviation between constant volume (data [E]) and constant pressure (data [C]) due to the small difference in the dimensionless stiffness as shown in the legend.

Figure 6b shows the same data but only the soft particle simulations ([C] and [E]) with pressure nondimensionalized by the particle stiffness. In this way, we observe a data collapse for high volume fractions, $\nu > \nu_c$, in agreement with the rate independent behavior as observed in other studies. This collapse of data for $\nu > \nu_c$ indicates that above the critical volume fraction the steady state rheological behavior of soft granular media under shear is dominated mostly by particle stiffness, while the influences of polydispersity and restitution coefficient ($e_n = 0.8$ in data [C] and $e_n = 0.7$ in data [E]) are of minor importance. In this regime, the higher the volume fraction the more solid like the behavior, and hence the less influences come from other microparameters than stiffness. It is also important to mention that even though we presented the analysis for pressure only, the shear stress shows a similar quantitative behavior [49].

4.6. So much for the granular rheology

While up to now, the focus was on understanding the relation between pressure and volume fraction, a granular rheology also must consider the shear stress.

Figure 7 shows the steady state shear stress ratio, $\mu = \tau/p$ (scaled by pressure, mostly referred as macroscopic friction), against inertial number for all the data discussed from **Figure 6a** (with different polydispersity, restitution coefficient, particle stiffness, as simulated in diverse numerical setups). It is important to realize that though both shear stress and pressure diverge close to the critical volume fraction point, their ratio does not. We observe the traditional μ(I)-rheology as a basic trend. For low inertial number, μ is almost independent of, I, and increases with increasing, I, for intermediate to large, I. Interestingly, although the qualitative trend of all the data is predicted by the traditional rheology, we still observe the deviations from the prediction in **Figure 7**. There are still many unveiled folders in the granular rheology like nonlocal behaviors, small shear rates diffusion, particle softness influence, etc., not to mention the complexity of including the frictional and cohesive granular media or/and with liquid

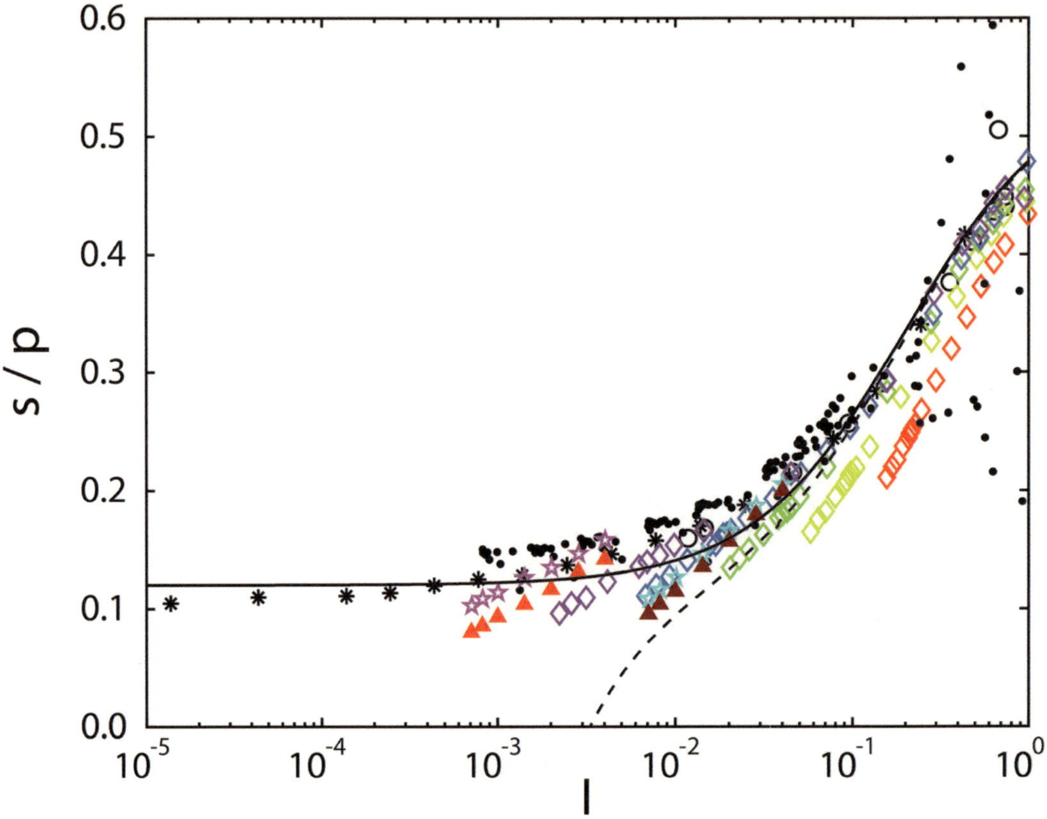

Figure 7. Steady state stress ratio (shear stress divided by pressure) versus inertial number, I, for data from different numerical setups as introduced in the legend of **Figure 6**, with different polydispersity, restitution coefficient, and particle stiffness. The black solid line shows the traditional μ(I)-rheology from Eq. (4) with the fitting parameters $\mu_0 = 0.12$, $\mu_\infty = 0.55$, and $I_0 = 0.2$ for frictionless rigid particles (black symbols) [64]. The dashed line represents the prediction of the extended rheology from Eq. (5) using (data [E]) with $p_0^* = 0.9$ [49].

bridges and suspensions. And also, the missing link between the dilute and dense granular rheological models is still a great challenge in the future.

5. Conclusion

This chapter gives an overview of recent progress in understanding and theoretically describing the collective mechanical behavior of dissipative, deformable particles in different states, both fluid-like and solid-like. Particulate systems and granular matter display collisional, dilute and solid, mechanically stable states, either switching forth and back, or both at the same time. In which state the system resides depends not only on material properties like, for example, their discrete nature (elastic stiffness), the dissipation (restitution coefficient) or the size distribution (polydispersity) of the particles, but also on the density of the system and balance between the energy input by (shear) stress or strain-rate and the energy dissipation by

http://dx.doi.org/10.5772/intechopen.68465

collisions or plastic deformations. Realistic material properties like friction and cohesion as well as nonsphericals particles go beyond the scope of this chapter.

One extreme case of low and moderate density collisional flows (for weak to moderate dissipation and arbitrary polydispersity) is well described by standard kinetic theory (SKT) up to system volume fractions about 0.5, beyond which the elastic behavior of longer-lasting contacts becomes dominant. Open challenges involve very soft particles for which basic theoretical assumptions of kinetic theory fail, for example, due to multiple contacting particles.

The other extreme case of quasi-static flow of elastic, mechanically stable solid-like structures are approximately described by the classical $\mu(I)$-rheology in the limit of rigid particles, but require a softness correction for comparatively large confining stresses. Remarkably, dissipation, as quantified by the coefficient of restitution, dominates the collisional flows in the dilute regime, while the particle stiffness, the polydispersity, and the friction (data not shown here) are the controlling microparameters for denser quasi-static and jammed flows.

The mystery of bridging the gap between the collisional, dilute, and the denser quasi-static, elastic solid-like regimes is not completely solved yet. The particulate, microscopic states are well understood by particle simulations that via so-called micro-macro transition can guide the development of macroscopic, continuum constitutive relations that allow to predict the state and characteristics where a granular system resides in. A unified description that ranges from dilute to dense, from rapid to slow, from soft to rigid, etc., is still one of the great challenges of today's research.

This chapter provided a few methods and some phenomenology, as well as an overview of recent literature in this field, with theories that can describe the extremes. Various recent works attempted to combine those limit-cases and provide first combined, generalized theories that go beyond the classical states. However, due to dissipation, friction, cohesion, and nonsphericity of realistic materials, this poses still plenty of challenges for today's research. Our own ongoing research focuses on providing simple unified/generalized theories, also for systems with attractive forces and with anisotropic microstructures, which were not addressed in this chapter.

Author details

Hao Shi[1]*, Dalila Vescovi[2], Abhinendra Singh[3], Sudeshna Roy[1], Vanessa Magnanimo[1] and Stefan Luding[1]

*Address all correspondence to: h.shi-1@utwente.nl

1 Multi Scale Mechanics (MSM), Engineering Technology (ET), MESA+, University of Twente Enschede, the Netherlands

2 Department of Civil and Environmental Engineering, Politecnico di Milano, Milano, Italy

3 Benjamin Levich Institute, City College of New York, New York, USA

References

[1] Göncü F, Luding S. Effect of particle friction and polydispersity on the macroscopic stress–strain relations of granular materials. Acta Geotechnica. 2013;**8**(6):629-643

[2] Kumar N, Luding S, Magnanimo V. Macroscopic model with anisotropy based on micro–macro information. Acta Mechanica. 2014;**225**(8):2319-2343

[3] Singh A. Micro-macro and rheology in sheared granular matter [PhD thesis]. The Netherlands: Universiteit Twente; 2014.

[4] Jenkins J, Richman M. Grad's 13-moment system for a dense gas of inelastic spheres. In: The Breadth and Depth of Continuum Mechanics. Berlin, Heidelberg: Springer, 1986. pp. 647-669

[5] Goldhirsch I. Stress, stress asymmetry and couple stress: From discrete particles to continuous fields. Granular Matter. 2010;**12**(3):239-252

[6] Goldhirsch I. Rapid granular flows. Annual Review of Fluid Mechanics. 2003;**35**(1):267-293

[7] Jenkins JT, Zhang C. Kinetic theory for identical, frictional, nearly elastic spheres. Physics of Fluids. 2002;**14**(3):1228-1235

[8] Bagnold RA, editor. Experiments on a gravity-free dispersion of large solid spheres in a Newtonian fluid under shear. In: Proceedings of the Royal Society of London A: Mathematical, Physical and Engineering Sciences; 1954: The Royal Society.

[9] Da CF, Emam S, Prochnow M, Roux J-N, Chevoir F. Rheophysics of dense granular materials: Discrete simulation of plane shear flows. Physical Review E. 2005;**72**(2):021309

[10] Lois G, Lemaître A, Carlson JM. Numerical tests of constitutive laws for dense granular flows. Physical Review E. 2005;**72**(5):051303

[11] Pouliquen O. Scaling laws in granular flows down rough inclined planes. Physics of Fluids. 1999;**11**(3):542-548

[12] Silbert LE, Ertas D, Grest GS, Halsey TC, Levine D, Plimpton SJ. Granular flow down an inclined plane: Bagnold scaling and rheology. Physical Review E. 2001;**64**(5):051302

[13] di Prisco C, Imposimato S. Experimental analysis and theoretical interpretation of triaxial load controlled loose sand specimen collapses. Mechanics of Cohesive, Frictional Materials. 1997;**2**(2):93-120

[14] Kolymbas D. Constitutive modelling of granular materials. Berlin, Heidelberg: Springer; 2012.

[15] Perzyna P. Fundamental problems in viscoplasticity. Advances in Applied Mechanics. 1966;**9**:243-377

[16] Roux J-N, Combe G. Quasistatic rheology and the origins of strain. Comptes Rendus Physique. 2002;**3**(2):131-140

[17] Wood DM. Geotechnical modelling. London and New York: Spon Press; 2004.

[18] Schofield A, Wroth P. Critical state soil mechanics. New York: McGraw-Hill; 1968.

[19] Campbell CS. Granular shear flows at the elastic limit. Journal of Fluid Mechanics. 2002;**465**: 261-291

[20] Ji S, Shen HH. Characteristics of temporalspatial parameters in quasisolid-fluid phase transition of granular materials. Chinese Science Bulletin. 2006;**51**(6):646-654

[21] Ji S, Shen HH. Internal parameters and regime map for soft polydispersed granular materials. Journal of Rheology. 2008;**52**(1):87-103

[22] Chialvo S, Sun J, Sundaresan S. Bridging the rheology of granular flows in three regimes. Physical Review E. 2012;**85**(2):021305

[23] MiDi G. On dense granular flows. The European Physical Journal E. 2004;**14**(4):341-365

[24] Pouliquen O, Cassar C, Jop P, Forterre Y, Nicolas M. Flow of dense granular material: Towards simple constitutive laws. Journal of Statistical Mechanics: Theory and Experiment. 2006;**2006**(07):P07020

[25] Gray J, Edwards A. A depth-averaged-rheology for shallow granular free-surface flows. Journal of Fluid Mechanics. 2014;**755**:503-534

[26] Berger N, Azéma E, Douce J-F, Radjai F. Scaling behaviour of cohesive granular flows. EPL (Europhysics Letters). 2016;**112**(6):64004

[27] Daniel RC, Poloski AP, Sáez AE. A continuum constitutive model for cohesionless granular flows. Chemical Engineering Science. 2007;**62**(5):1343-1350

[28] Jop P, Forterre Y, Pouliquen O. A constitutive law for dense granular flows. Nature. 2006;**441**(7094):727-730

[29] Jop P. Hydrodynamic modeling of granular flows in a modified Couette cell. Physical Review E. 2008;77(3):032301

[30] Singh A, Magnanimo V, Saitoh K, Luding S. The role of gravity or pressure and contact stiffness in granular rheology. New Journal of Physics. 2015;**17**(4):043028

[31] Roy S, Singh A, Luding S, Weinhart T. Micro–macro transition and simplified contact models for wet granular materials. Computational Particle Mechanics. 2016;**3**(4):449-462

[32] Roy S, Luding S, Weinhart T. A general(ized) local rheology for wet granular materials. New Journal of Physics. 2017;**19**:043014.

[33] Brilliantov NV, Pöschel T. Kinetic theory of granular gases. New York: Oxford University Press; 2004.

[34] Garzó V, Dufty JW, Hrenya CM. Enskog theory for polydisperse granular mixtures. I. Navier-Stokes order transport. Physical Review E. 2007;**76**(3):031303

[35] Garzó V, Hrenya CM, Dufty JW. Enskog theory for polydisperse granular mixtures. II. Sonine polynomial approximation. Physical Review E. 2007;**76**(3):031304

[36] Goldhirsch I. Introduction to granular temperature. Powder Technology. 2008;**182**(2):130-136

[37] Pöschel T, Luding S. Granular gases. Berlin, Heidelberg: Springer; 2001.

[38] Goldhirsch I, Sela N. Origin of normal stress differences in rapid granular flows. Physical Review E. 1996;**54**(4):4458

[39] Garzó V, Dufty J. Dense fluid transport for inelastic hard spheres. Physical Review E. 1999;**59**(5):5895

[40] Alam M, Willits JT, Arnarson BÖ, Luding S. Kinetic theory of a binary mixture of nearly elastic disks with size and mass disparity. Physics of Fluids. 2002;**14**(11):4085-4087

[41] Berzi D, Vescovi D. Different singularities in the functions of extended kinetic theory at the origin of the yield stress in granular flows. Physics of Fluids. 2015;**27**(1):013302

[42] Goldshtein A, Shapiro M. Mechanics of collisional motion of granular materials. Part 1. General hydrodynamic equations. Journal of Fluid Mechanics. 1995;**282**:75-114

[43] Lun C. Kinetic theory for granular flow of dense, slightly inelastic, slightly rough spheres. Journal of Fluid Mechanics. 1991;**233**:539-559

[44] Lun C, Savage S. A simple kinetic theory for granular flow of rough, inelastic, spherical particles. Journal of Applied Mechanics. 1987;**54**(1):47-53

[45] Pöschel T, Buchholtz V. Molecular dynamics of arbitrarily shaped granular particles. Journal de physique I. 1995;**5**(11):1431-55

[46] Garzó V, Dufty JW. Hydrodynamics for a granular binary mixture at low density. Physics of Fluids. 2002;**14**(4):1476-1490

[47] Jenkins JT. Dense shearing flows of inelastic disks. Physics of Fluids. 2006;**18**(10):103307

[48] Jenkins JT. Dense inclined flows of inelastic spheres. Granular Matter. 2007;**10**(1):47-52

[49] Vescovi D, Luding S. Merging fluid and solid granular behavior. Soft Matter. 2016;**12**(41): 8616-8628

[50] Rognon PG, Roux J-N, Naaïm M, Chevoir F. Dense flows of bidisperse assemblies of disks down an inclined plane. Physics of Fluids. 2007;**19**(5):058101

[51] Kamrin K, Koval G. Nonlocal constitutive relation for steady granular flow. Physical Review Letters. 2012;**108**(17):178301

[52] Luding S, Singh A, Roy S, Vescovi D, Weinhart T, Magnanimo V, editors. From particles in steady state shear bands via micro-macro to macroscopic rheology laws. In: Li X, Feng Y, Mustoe G (eds) Proceedings of the 7th International Conference on Discrete Element Methods DEM 2016 Springer Proceedings in Physics; 2017. Singapore: Springer.

[53] Cundall PA, Strack OD. A discrete numerical model for granular assemblies. Geotechnique. 1979;**29**(1):47-65

[54] Hartkamp R, Todd B, Luding S. A constitutive framework for the non-Newtonian pressure tensor of a simple fluid under planar flows. The Journal of Chemical Physics. 2013;**138**(24):244508

[55] Hartkamp R, Ghosh A, Weinhart T, Luding S. A study of the anisotropy of stress in a fluid confined in a nanochannel. The Journal of Chemical Physics. 2012;**137**(4):044711

[56] González S, Thornton AR, Luding S. An event-driven algorithm for fractal cluster formation. Computer Physics Communications. 2011;**182**(9):1842-1845.

[57] Radjai F, Richefeu V. Contact dynamics as a nonsmooth discrete element method. Mechanics of Materials. 2009;**41**(6):715-728

[58] Luding S. Introduction to discrete element methods: Basic of contact force models and how to perform the micro-macro transition to continuum theory. European Journal of Environmental and Civil Engineering. 2008;**12**(7-8):785-826

[59] Shäfer J, Dippel S, Wolf D. Force schemes in simulations of granular materials. Journal de physique I. 1996;**6**(1):5-20

[60] Weinhart T, Thornton AR, Luding S, Bokhove O. From discrete particles to continuum fields near a boundary. Granular Matter. 2012;**14**(2):289-294

[61] Shi H, Luding S, Magnanimo V. Limestone Powders Yielding and Steady State Resistance under shearing with different testers. in: 2nd International Conference on Powder, Granule and Bulk, 2016 (PGBSIA 2016), India, Jaipur. 2016.

[62] Weinhart T, Hartkamp R, Thornton AR, Luding S. Coarse-grained local and objective continuum description of three-dimensional granular flows down an inclined surface. Physics of Fluids. 2013;**25**(7):070605

[63] Singh A, Magnanimo V, Saitoh K, Luding S. Effect of cohesion on shear banding in quasistatic granular material. Physical Review E. 2014;**90**(2):022202

[64] Peyneau P-E, Roux J-N. Frictionless bead packs have macroscopic friction, but no dilatancy. Physical Review E. 2008;**78**(1):011307

[65] Chialvo S, Sundaresan S. A modified kinetic theory for frictional granular flows in dense and dilute regimes. Physics of Fluids. 2013;**25**(7):070603

[66] Ogarko V, Luding S. Equation of state and jamming density for equivalent bi-and polydisperse, smooth, hard sphere systems. The Journal of Chemical Physics. 2012;**136**(12): 124508

[67] Ogarko V, Luding S. Prediction of polydisperse hard-sphere mixture behavior using tridisperse systems. Soft Matter. 2013;**9**(40):9530-9534

[68] Kumar N, Imole OI, Magnanimo V, Luding S. Effects of polydispersity on the micro–macro behavior of granular assemblies under different deformation paths. Particuology. 2014;**12**:64-79

Section 3

Flow State of Granular Materials

Chapter 4

Dynamics of a 2D Vibrated Model Granular Gas in Microgravity

Yan Grasselli, Georges Bossis, Alain Meunier and
Olga Volkova

Additional information is available at the end of the chapter

http://dx.doi.org/10.5772/intechopen.68277

Abstract

We are reporting an experimental study performed on a granular gas enclosed into a 2D cell submitted to controlled external vibrations. Experiments are performed in microgravity during parabolic flights. High-speed optical tracking allows to obtain the kinematics of the particles and the determination of all inelastic parameters as well as the translational and rotational velocity distributions. The energy into the medium is injected by submitting the experimental cell to an external and controlled vibration. Two model gases are studied beads and disks; the latter being used to study the rotational part of the particle's dynamics. We report that the free cooling of a granular medium can be predicted if we consider the velocity dependence of the normal restitution coefficient and that the experimental ratio of translational versus rotational temperature decreases with the density of the medium but increases with the driving velocity of the cell. These experimental results are compared with existing theories. We also introduce a model that fairly predicts the equilibrium temperatures along the direction of vibration.

Keywords: granular, microgravity, translational temperature, rotational temperature

1. Introduction

Granular gases are suspensions in air of macroscopic particles whose dynamics is ruled by momentum transfer during collisions between the particles. Unlike molecular gases, these collisions are not elastic, and the dissipation resulting of each collision gives important qualitative differences, since without bringing energy to the system, the motion of the particles will quickly stop. The supply of energy can be natural, as it is the case with gravity forces during avalanches or due to a flow of fluid through a bed of particles, or artificial for instance by shaking a box containing the grains. Due to the importance of granular flows in many industries, they have been the subject of intensive research and numerous reviews [1–5].

Gravity is, of course, a fundamental parameter, which governs the density distribution of particles with height in a sheared flow or in a vibrated container. The understanding of the complex physics of these granular systems is then complicated by the presence of the gravity. Besides, numerical simulations allow to obtain information on the dynamics of model systems of granular particles without the need to use sophisticated experiments in parabolic flights, drop tower or suborbital rocket flight but can't replace real experiments [6–9].

Two specific phenomena of the dynamics of a vibrated granular are the clustering instabilities, which occur due to the dissipation in multiple collisions between grains and the violation of the equipartition energy between each translational and rotational degree of freedom.

Recent results, obtained by molecular dynamics simulations of a box with two opposite vibrating walls and fixed side walls, have shown that in zero gravity the cluster of particles oscillate around its equilibrium position [10]. Experiments made in a parabolic flight in similar conditions with two opposite vibrating walls and with two different sizes of particles (diameter of 1 and 2 mm) were compared to simulation results. A phase diagram of clustering versus the volume fraction of each species obtained by numerical simulation was well agreeing with the experimental results and showed a segregation effect [11]. A different kind of cluster consisting of regular alignments of particles along the velocity lines in a Couette flow was also found in parabolic flight experiments [12].

Concerning the temperature of a vibrated granular gas, it was shown to follow a power law: $T = CV_p^\alpha$, where V_p is the peak amplitude of the vibration velocity and $1 < \alpha < 2$. In a recent paper [13], it was demonstrated that the different values of the power α were related to the ratio, W, of the energy injected by the vibration mV_p^2 to the gravitational potential energy. When W is large, that is, to say when the gravity is negligible, the value $\alpha = 2$ is recovered. A balance of the energy flux injected by the vibrated wall with the dissipation induced by particle-particle and particle-wall collisions allows to demonstrate that the temperature along the direction of vibration should be larger than the transverse temperature and that this ratio increases when the radial restitution coefficient decreases [43]. It was also found theoretically [38] that, even if the temperature of translational and rotational degrees of freedom were initially identical, the decrease of the translational temperature, after switching off the energy supply, was much faster than the rotational one but both of them are predicted to have a decrease in t^{-2} as predicted by Haff's law, although at longer time, the system becomes inhomogeneous and follows a decrease with $t^{-6/5}$ [14]. The t^{-2} decrease and the difference between translational and rotational energies were also obtained by numerical simulations for particles with a needle shape [15]. In simulations of the cooling of a gas of ellipsoidal grains with different aspect ratio [16], the authors have also found a t^{-2} law in all cases, but when the aspect ratio (a/b) increases the difference between the translational and rotational temperature decreases and had totally disappeared for a/b = 2, even becoming slightly larger than one for larger values of a/b. This is explained by the fact that the coupling between translational and rotational velocity is strongly increased by the shape's anisotropy. Also, they do not observe significant deviation from the Gaussian distribution for the velocity distribution. On the other hand, experiments in a rocket flight with a box having three moveable walls and containing needles of aspect ratio close to 10 [17] show a non-Gaussian distribution of the velocities in the excitation direction. In this experiment, the temperature perpendicular to the excitation direction as well as the rotational temperature was approximately two times less than the temperature in the excitation direction. In this last experiment, the positions of the particles were determined with the help of two video cameras at right angle, and the determination of the positions and orientations was done manually from each video frame. Although the physics of 2D and 3D system can differ in some aspects, the use of cells where particles are confined in 2D allows a much easier tracking of the particles and an automatic detection of their position. Such experiments made in low gravity with a determination of the trajectories of each particle are scarce but very useful to check the validity of models and of numerical simulations. This is the case for instance with the confirmation of the difference of the temperatures in the directions parallel and perpendicular to the external vibration [18] and for the comparison of the cooling time with the theoretical expression [19] where the experiment gave a time of the same order of magnitude as the theory: $\tau_{exp} = 38\ ms$ against $\tau_{th} = 60\ ms$ [20].

Two model systems made of inelastic hard spheres or disks will be used as reference models to study the dynamics. The general experimental situation is to have the particles enclosed in a vibrated box (for energy input) where the vibration parameters (amplitude and frequency) are monitored. Direct optical observations can lead to the dynamics of individual particle and to retrieve the physical data. The collisions between particles are leading the dynamics of the system through the inelastic interactions and momentum transfer. The normal and tangential restitution coefficients depend on the material of the particles but sometimes also on the impact velocity. Gravity being one of the main issues to overcome when studying a granular medium, the experimental results presented here have been performed in a low-gravity environment. Experiments were boarded in the Airbus Zero-G from Novespace (www.novespace.fr), and the results presented obtained during parabolic flights. 2D cells containing the granular particles were mounted on a vibrated device and high-speed video recording was used to register and track the motion of each individual particles.

We will first report an experimental study on the free cooling of a granular medium made of beads: focusing on the time relaxation of the energy of the medium, and then, we will present similar experiments realized with disks in order to get access to the granular temperatures for the translation and rotational part of the particle's energy.

2. Free cooling

The cooling of a granular gas can be experimentally investigated by considering a granular medium submitted to a continuous external energy input (generally done by submitting the medium to a controlled vibration), then removing it and observing how the medium goes back to rest. Experimental studies on granular have to deal with gravity effects and studying model particles (disks in general) on an air flow table can overcome gravitational effects. Our approach was to perform experiments in a low-gravity environment by boarding the experimental apparatus in the Airbus Zero-G from Novespace. The airplane undergoes successive parabolic flights allowing around 22 s of microgravity per parabola. The relative gravity is recorded during the flight (**Figure 1**) allowing to monitor the quality of microgravity environment.

Irons beads with radius a = 1 mm enclosed in a 2D cell (**Figure 2**) have been used as model granular particles. The area fraction of the medium was $\phi_i = 19\%$. The cell was chosen with a circular geometry to ensure a homogeneous energy input into the system while submitted to the vibration. The walls of the cell are made of glass to cancel as much as possible electrostatic effects and to reduce undesired friction effects between the particles and the walls. The cell is mounted on a vibrating device ("Modal exciter, 100N, Bruel&Kjaer") to allow periodic (sine oscillations) external vibrations with different frequencies, ν, and amplitudes A. The maximum cell's velocities can be changed from 30 cm/s up to 250 cm/s. This experimental set up prevents us from density fluctuations found in fluidized beds or strong rolling contributions encountered when particles move over a horizontal vibrated plate. The motion of granular particles is recorded with a high-speed camera at 470 fps during 6 s in each experiment. About 3000 pictures (320 × 320 pixels) can be retrieved from each recording.

The high-speed video recording allows us to track each particle and then to get access to their positions inside the cell. From this knowledge, the dynamics of the medium can be retrieved through the velocities of each particle. The experimental processing is performed by image analysis [21]. Each particle p is tracked individually allowing to obtain the positions $x_p(t)$ and $y_p(t)$ as a function of time. It is interesting to note that from these sets of coordinates all experimental parameters required to describe the collective motion can be directly determined such as the velocities components, the normal restitution coefficient, e, but also the pair correlation function g(r) (**Figure 3**). The maximum of g(r) is found at the particle diameter

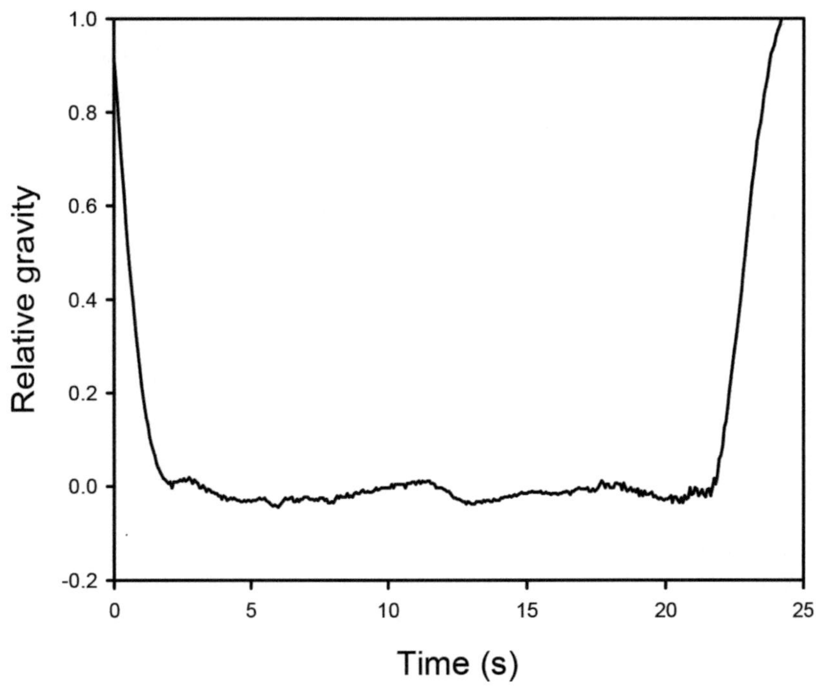

Figure 1. Typical behavior of the relative gravity during a parabola. These curves are used to check the quality of the microgravity environment.

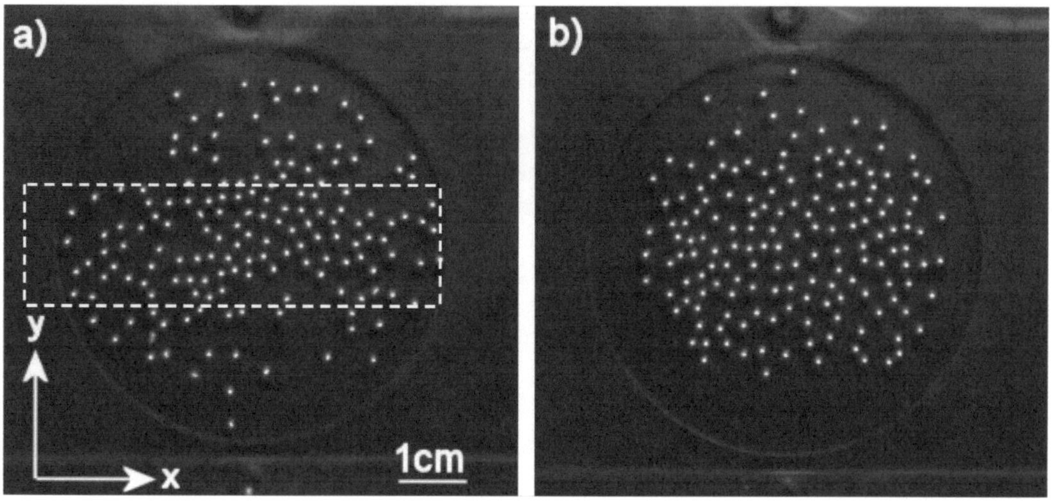

Figure 2. Snapshots of the experimental cell. The external vibration is applied along the y-direction (direction of the normal gravity). High-speed video recording is used to track the motion of each individual particle. (a) Vibration is on: the central part of cell contains an almost constant density of particles (dashed region). (b) The external vibration has been stopped and the overall motion of particles stops due to the inelastic collisions.

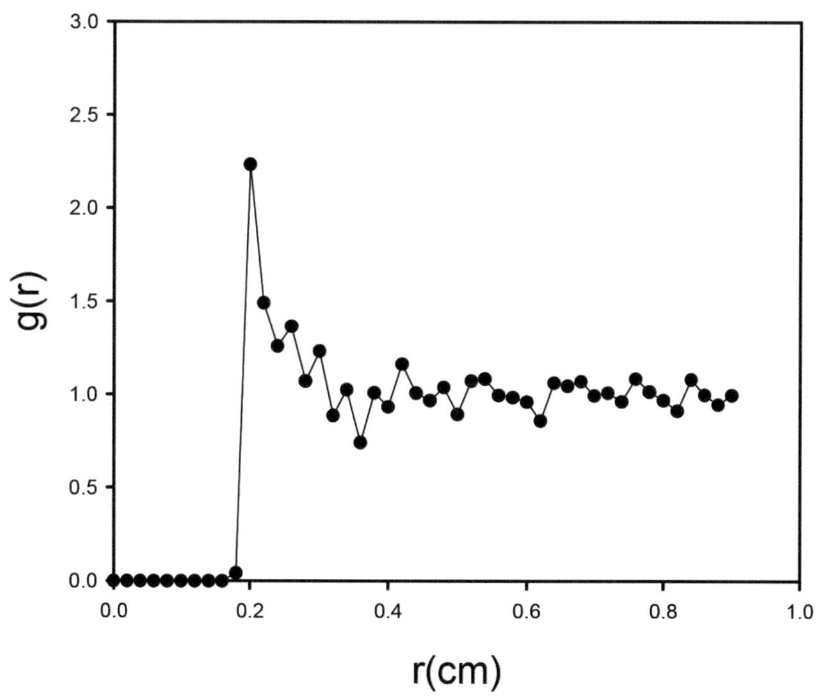

Figure 3. Experimental pair correlation function g(r) retrieved from the positions of the particles. This curve is averaged over all pictures recorded and on the spatial configurations of particles in the central area of the cell.

which proves that electrostatic effects are negligible. The small non-null value of the pair correlation function observed "before" the particle diameter is due to the uncertainty in the particle's position by image treatment.

In order to study the free cooling, that is, to relate the loss of energy of the medium due to the inelastic collisions between particles, the external vibration is switched on prior the micro-gravity occurs. In zero-g, the particles will then fill the entire region of the cell, and the video recording is started. After few seconds, the external vibration is switched off, and we observe the return to equilibrium (particles at rest throughout the cell). It is worth noting that in the presence of the vibration, two types of different regions clearly appear in the cell: two hot (and dilute) regions at the top and bottom of the cell while a dense region exists in the center of the cell (dashed area shown in **Figure 2a**). This experimental configuration gives us the possibility to study a homogeneous bed of particles in contact with two hot regions responsible for the energy input. As the external vibration is cancelled, the particles continue to move freely throughout the cell and tend to come to rest rapidly because of inelastic collisions between particles inducing energy loss. For cooled granular media, the formation of dense clusters of particles is often reported in experiments but it is not clearly observed in our situation: we rather observe some alignments of particles along "wavy lines" but there is no evidence of high and low-density regions as the main part of the energy loss is supposed to occur along the normal direction between two particles. The relative low area fraction of particles is also a possible reason for this non-observation of this clustering effect. Moreover, g-jitter still exists could add a general motion of particles in a given direction. But a short

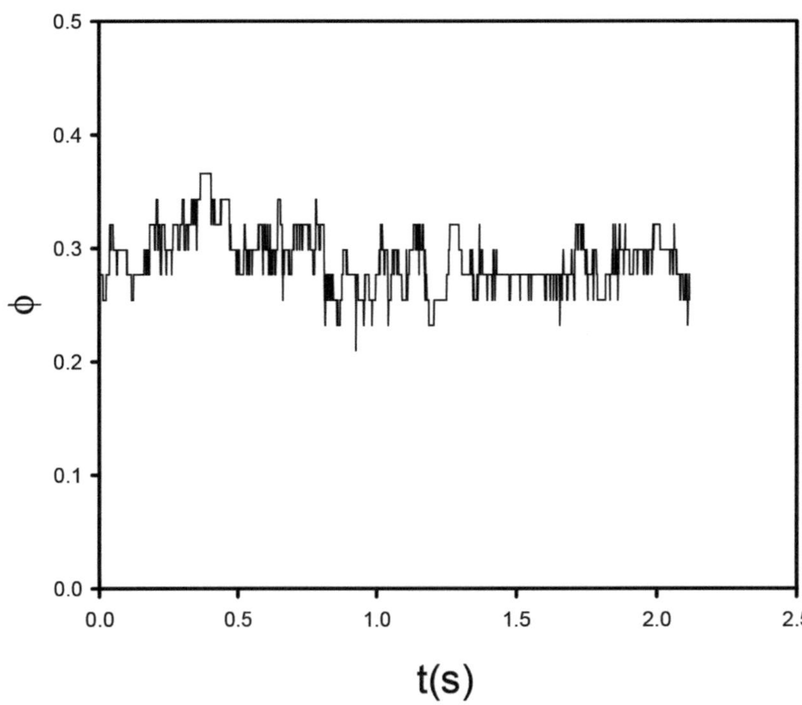

Figure 4. Volume fraction of particles in the central area of the cell as a function of the recording time (see **Figure 2a**). In this region, we will assume that the volume fraction of the granular medium remains constant.

http://dx.doi.org/10.5772/intechopen.68277

time, after the vibration has been removed, we generally observe that the particles tend to stop in the center of the cell without evidence for clustering. During the recording, we have verified that the concentration in the central part of cell remains constant, and we have based all of other study on the dynamics of this area (**Figure 4**).

As the behavior of the medium is governed by inelastic collisions, we have determined experimentally the normal restitution coefficient as a function of the relative normal velocities of two colliding particles. A systematic investigation of binary collisions of particles has been realized either in the presence of the external vibration or without it. From the optical tracking and the knowledge of the trajectories, we can compute the directions and the magnitudes of the velocities before, $\overrightarrow{V_R}$ and after, $\overrightarrow{V'_R}$ an impact between particles. By tracking these changes in the direction of motion of each particle when a nearest neighbor is present, we are able to precisely determine the binary collisions from the trajectories and so the exact position of the colliding particles. It is then possible to consider the positions of particles around the location of the collision (**Figure 5**) and insure that the trajectories before and after collision are linear to

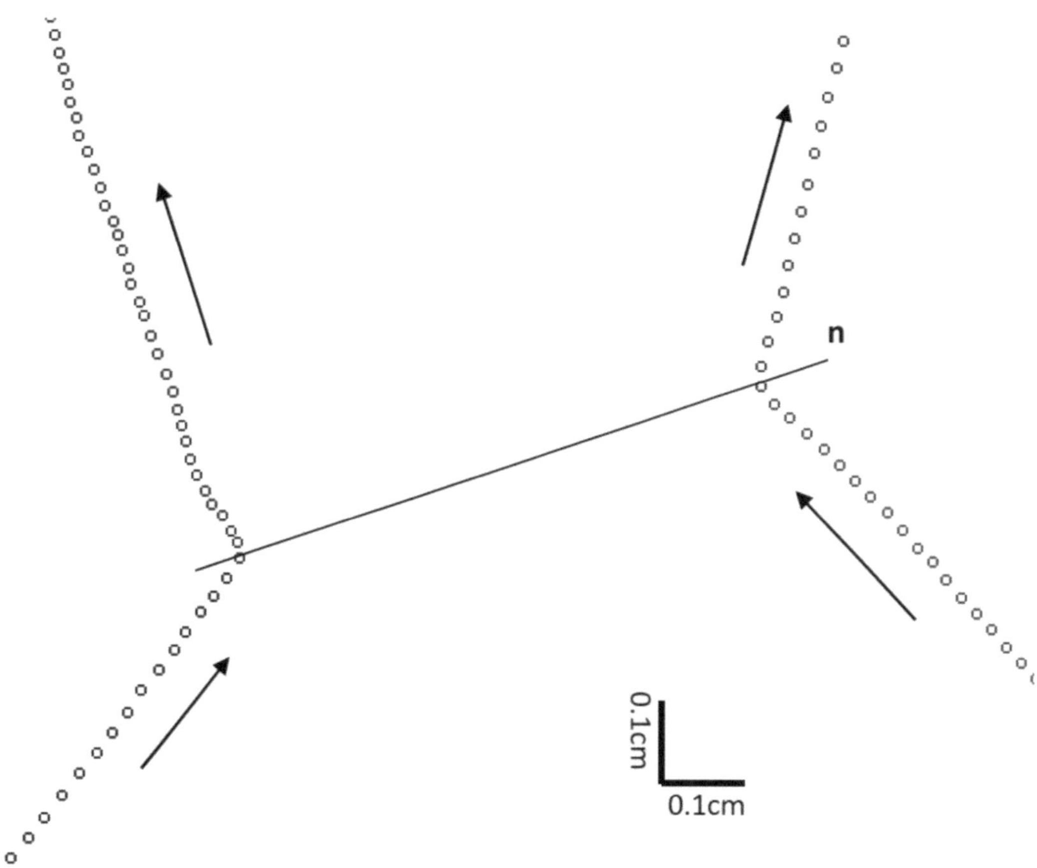

Figure 5. Experimental trajectories recorded during a binary collision between particles. The circles represent the positions retrieved from optical tracking. For a better understanding, we have added on the experimental trajectories the direction of motion (arrows) before and after collision. We can precisely obtain the position of each particle at impact and the direction of the normal direction n.

qualify this collision for processing. The direction of the normal direction at contact is then possible, and the restitution coefficient is obtained from $e = -| \vec{n} \cdot \overrightarrow{V'_R} |/| \vec{n} \cdot \overrightarrow{V_R}|$.

The behavior of the restitution coefficient vs. the normal relative impact velocity is presented in **Figure 6**. For high relative velocities, we obtain a value of the restitution coefficient of 0.9 (typical value for steel beads). The most amazing observation is that the restitution coefficient shows a sharp decrease for "small" impact velocity. This is a situation encountered in the case of wet particles when e = 0 for Stokes number $S_t = (mV_i/6\pi\eta a2)$ smaller than a critical value [22, 23]. This comes from the viscous dissipation but for dry particles, most experimental investigations report that the restitution coefficient increases for increasing impact velocities. Most of the experiments are made in labs (i.e., with gravity present) with impact velocities larger than 1 m/s (for a height h = 5 cm the impact velocity of a bead on the plane: $V_i = \sqrt{2gh}$ is already 1 m/s). The restitution coefficient between two dry beads attached by strands in a pendulum device has also been studied [24], where it has also been found that a low value of the restitution coefficient was reported at low velocities (typically below 20 cm/s), and such behavior is well confirmed in our study without experimental drawbacks.

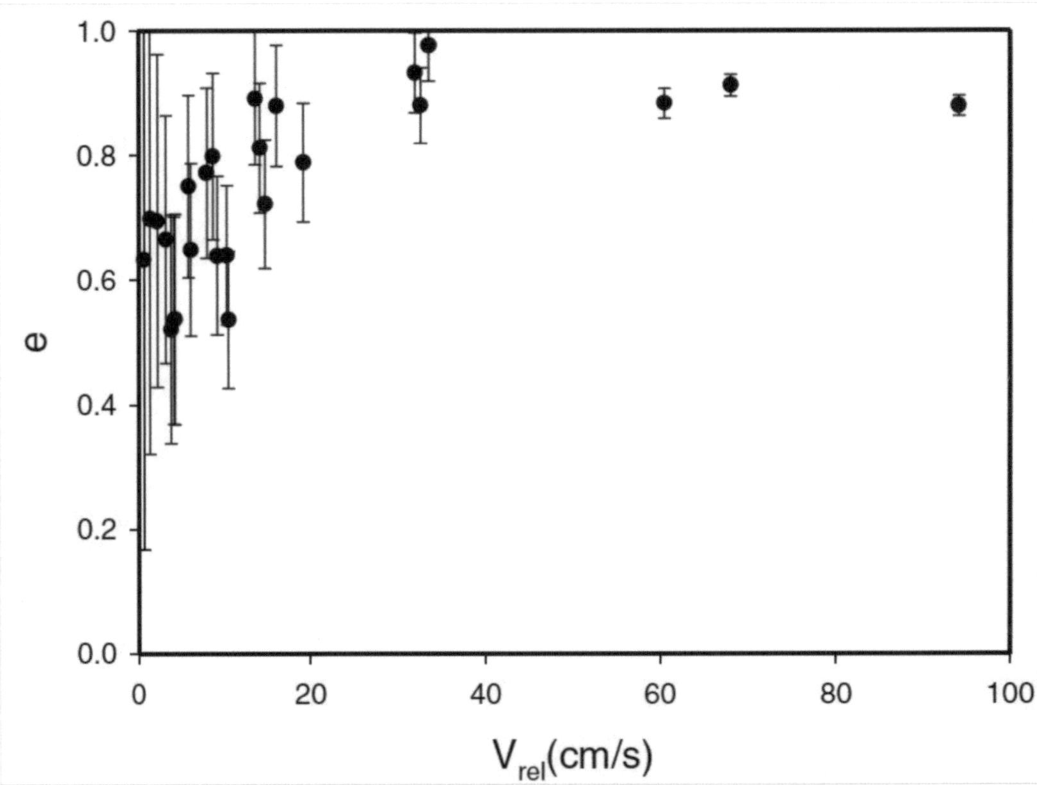

Figure 6. Experimental dependence of the normal restitution coefficient, e, as a function of the relative normal impact velocity, obtained from the experimental trajectories of the particles. A clear decrease at low impact velocities is observed.

To investigate the free cooling more precisely, a typical record on how the energy decreases once the external energy input has been cancelled is presented in **Figure 7**. This behavior is monitored through the average velocities of the particles in the central area of the cell. One can observe the rapid decay of the average velocity. The non-zero value measured for "long times" comes from the small gravity fluctuations occurring during the parabolic flight.

We can first consider the energy decay assuming a constant restitution coefficient (typically $e = 0.9$ for stainless steel beads). The time dependence of the energy is predicted to behave as $E(\tau) = 1/(1+\tau)^2$, $\tau = (1-e^2)t/t_E$, where t_E is the Enskog time [25]: $t_E = (a\sqrt{\pi})/\left(\sqrt{2}\phi V_0 g(r)\right)$. V_0 is the initial average velocity in the medium. With our experimental set up, we can determine experimentally all the parameters involved. A quantitative comparison with experiments is presented in **Figure 8** (squared symbols) for a cell velocity of 75 cm/s and with the following experimental values: $\phi = 0.297 \pm 0.027$, $V_0 = (0.11 \pm 0.01)\ m/s$ and $g(r = 2a) = 2.23 \pm 0.02$. The drop of energy found from experiments is much faster than the theoretical one while considering a constant restitution coefficient. A possible explanation may result from the friction of the particles on walls of the cell, introducing an additional loss of energy. Nevertheless, a precise and systematic analysis of the trajectory of a single particle after the vibration has been cut off shows a linear motion at constant speed between two collisions of particles; we may then reject this possibility. The Enskog collision time is $t_E = 0.0172\ s$. In order to check this value, we

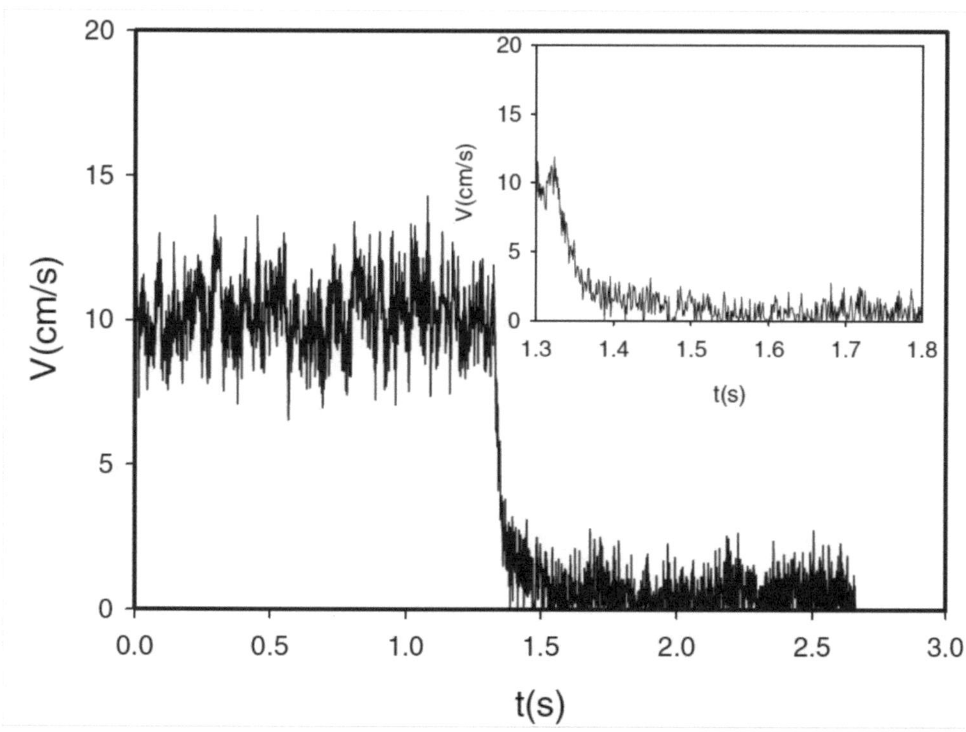

Figure 7. Average translational velocity as a function of time obtained in the central region of the cell. The vibration is removed during the microgravity period. A clear decrease of the energy can be observed (max cell velocity of 74.6cm/s).

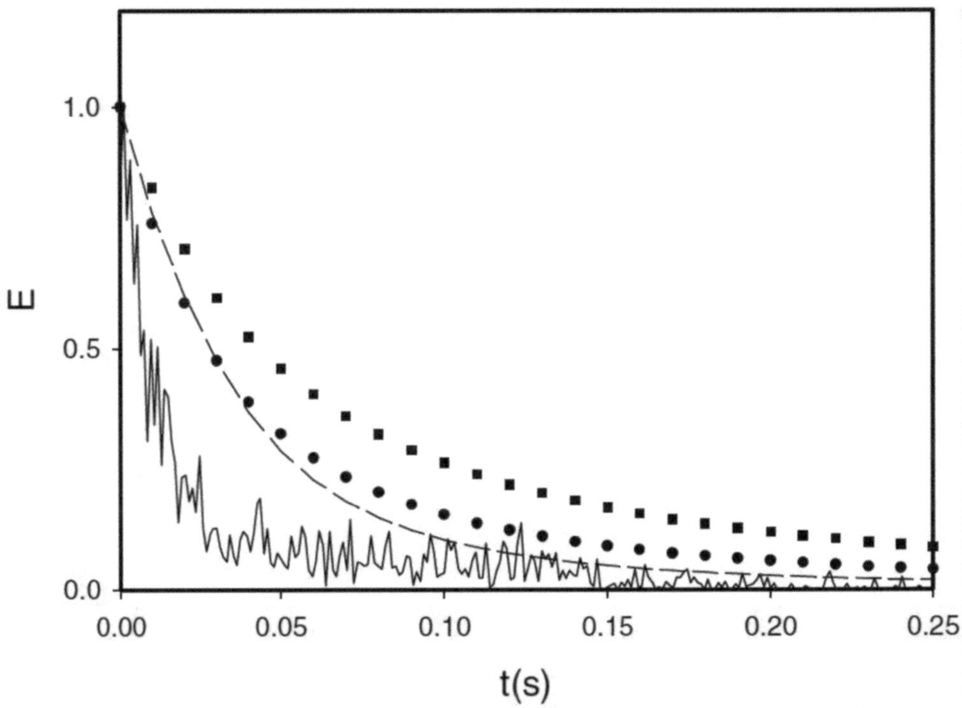

Figure 8. Experiments (plain curve) and theory of the energy decrease. Squares: theory including a constant restitution coefficient. Circles: theory considering the rotational energy. Dashed line: theory focusing only the translational energy but including the velocity dependence of the restitution coefficient (see **Figure 6**).

have performed a large statistics on our experiments to obtain the average time interval separating two consecutive collisions in the central part of the cell. We found an average time interval of (0.0127 ± 0.0021) s by direct measurements in rather good agreement with the theoretical value.

So, the possible discrepancy between energy decays observed experimentally and the predicted one by theory may come from the rotational kinetic energy which also dissipates a part of the energy through the surface roughness of the particles [26]. To get a complete description of the binary collision, we have introduced a tangential restitution coefficient, β, in order to characterize the contribution of the rotation of the particles. The time dependence of the translational and rotational energy is obtained from coupled differential equations (Eq. (15) in Ref. [26]): note that the parameters considered in this description are all retrieved from experiments, except β. This system of equations was solved numerically. We have introduced our experimental results for the inelastic parameters of particles and setting $\beta = 0.1$ (**Figure 8**, black circle—if we cancel the rotation, that is, $\beta = -1$, we recover the situation of a constant normal restitution coefficient). We see that the energy decreases more rapidly but it seems that the rotational kinetic energy has limited impact whatever the value of the tangential restitution coefficient and it is still not enough to represent the experimental behavior.

To improve the agreement between theory and experiments, we may consider the velocity dependence of the restitution coefficient. We can express the rate of decrease of the translational kinetic energy T like $dT/dt = -n_c(1 - e2)T$ with n_c the rate of binary collisions. In 2D, $n_c = \left(2V\phi g(r)\right)/(\pi a)$ with V, the average velocity. Moreover, introducing the normalized energy $E = T/T_0$ and the velocity ratio $V/V_0 = T/T_0$, the rate of decrease of energy can be rewritten in the form:

$$\frac{dE}{dt} = -\frac{g(r)\phi V_0}{a}\sqrt{\frac{2}{\pi}}\left(1 - e(E)^2\right)E^{3/2} \tag{1}$$

But now e is assumed to depend on the normal relative velocity. We may assume that the average relative velocity is of the same order as the average velocity; then from **Figure 6**, we can obtain the following trend $e(E) = 0.82 - 0.5e^{-2.5E}$. Substituting this last relation in Eq. (1) and solving it numerically gives the behavior presented in **Figure 8** (dashed line). Compared to the case including the rotation, we observe a more pronounced decrease of the energy with time. This is understandable since we observed that during cooling the restitution coefficient decreases, increasing the loss of energy. Equation (1) is obtained from a rough approximation based on the average velocity, while the probability distribution may be considered. Moreover, the tangential coefficient may probably also be dependent on the relative angular velocities of colliding particles.

This first approach on the dynamics of a granular medium shows interesting results but as stated before, the analysis is not complete due to the lack of consideration on rotational effects. With beads and our experimental setup, accessing these data is not possible. We will then introduce in the next part recent experimental investigations based on the same principle but replacing beads by disks in order to obtain a complete description of the dynamical behavior.

3. Translational and rotational temperatures

In this part, our aim is to provide experimental data both for the normal and tangential restitution coefficients and for the different quantities related to the rotational and translational degrees of freedom such as the distribution functions and the rotational and translational temperatures. As introduced previously, the kinematics of granular particles submitted to a vertical vibration will still be used in a low-gravity environment. We shall particularly focus on the ratio between rotational and translational temperatures. Several other groups have already presented experimental results on granular flow under such conditions [27–30], but to our knowledge, this is the first experimental study giving access to rotational and translational velocities and so the corresponding temperatures.

We have used the same 2D cell from the previous part by now with a rectangular shape made in Duralumin. The cell has a height $L_y = 6.8$ cm and a width $L_x = 6$ cm. The particles studied were brass disks having a diameter $\sigma = 6$ mm (radius $a = 3$ mm) and mass $m = 4.6{\cdot}10^{-4}$ kg. The initial area fraction ϕ of the medium is obtained from the number of disks N (12 or 24, i.e., area fractions of 8.3 or 16.6%). For this experimental study, we chose this rectangular shape in order to easily monitor the energy input into the medium. The external vibration is still periodic (sine oscillations) with different frequencies, ν, and amplitudes A, and it is still applied along the y-direction (which is the direction of normal gravity). In order to increase the precisions of experimental data, the video recording is performed during the whole parabola with a higher frame rate (i.e., 900 fps) and higher image resolution (720 × 720 Pixels); each record gives access to around 22,000 pictures per parabola. To reduce the effect of friction between the disks and the glass plates of the cell, we have added three small steel beads on each side of a disk. In addition, it also reduces the tilting of the disks when the external vibration is on. The key question being now the rotational aspects, each disk is pierced with two small holes, symmetric about the center of the disk and video observations are realized by light transmission (**Figure 9**).

This set up grants us with images having a high contrast and quality. The position of the disk is retrieved from the tracking of the two holes for each disk as a function of time. The barycenter of the holes gives access to the x- and y-position of the disk and by following the variations of these positions as a function of time to the components of the velocity $v_x(t)$ and $v_y(t)$. Nevertheless, by computing the time dependence of the angle $\theta(t)$ (**Figure 10**) (obtained from the angular position of the holes about the horizontal), allows to retrieved the angular velocity $\dot{\theta}(t)$.

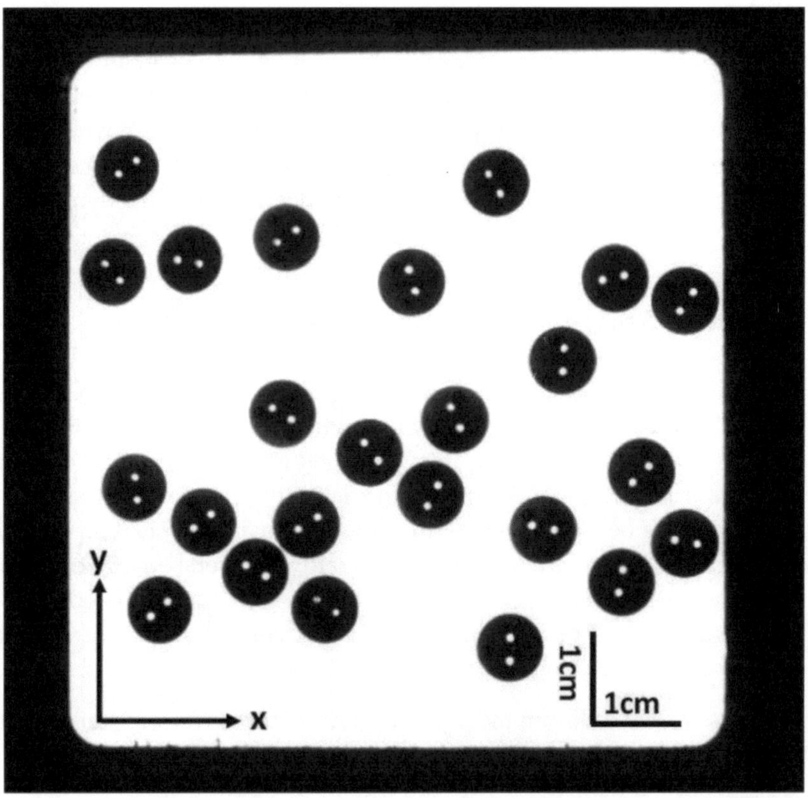

Figure 9. Picture of the medium recorded in microgravity when being submitted to the external vibration (along the y-direction). Optical observations are performed from light transmission. The two holes can be clearly identified. A side and top sketch of one disk is also shown.

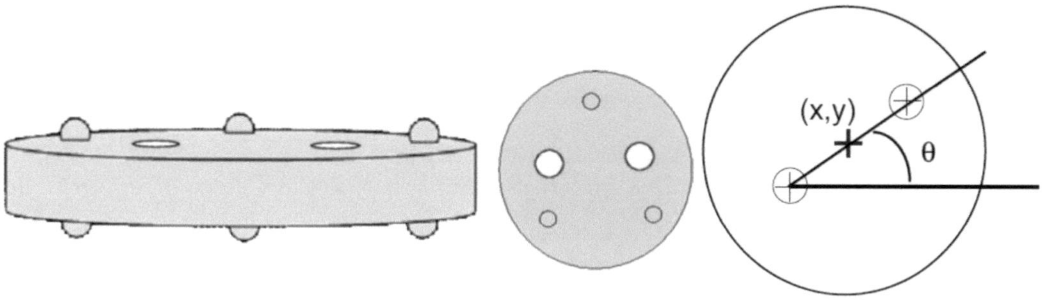

Figure 10. Disks used as the granular particles. Light transmission allows very high contrast pictures. The optical tracking of the two holes of a single disk permits to compute the orientation angle of the disk as a function of time.

The orientation angles of the disks can be fully determined from 0 to 360 degrees. A typical experimental record of θ(t) is presented in **Figure 11**. On such record, a sharp change in the direction of rotation (positive or negative slope) or a significant change in the slope is the proof that a collision occurs with another particle. On the contrary, when the particle experiences no collision (e.g., time larger than 5 s in **Figure 11**), the angular velocity remains quite constant, indicating the absence of friction with the lateral walls. During a parabolic flight, the aircraft is subjected to g-jitter along the three directions (**Figure 1**). Experiments were submitted to g-jitter with period of fluctuations of about 1 s and amplitudes of about 0.01 g (**Figure 1**). Although these fluctuations may play a role during the collision of the particles with the moving walls of the cell, they have limited impact on the motion of particles located in the central region of the cell where experimental data were retrieved. Moreover, a systematic

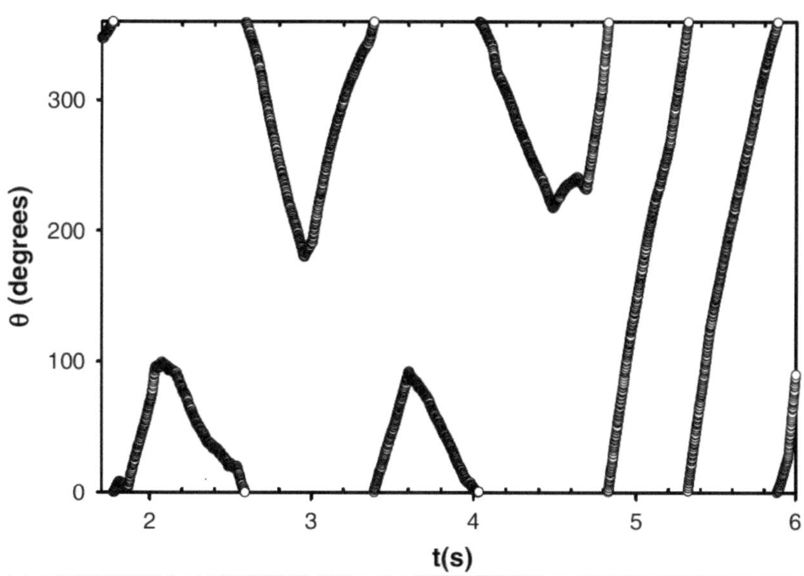

Figure 11. Experimental record of the angle of orientation, θ(t) of a disk when both microgravity and external vibration are present. Collisions can be clearly identified from a change of rotation or value of the angular velocity (i.e., The slope).

analysis of inelastic parameters (normal e, and tangential β, restitution coefficients) was achieved by analyzing the trajectory of each disk.

The collision between disks is processed as we did for the beads in the previous part except that now, the relative velocity includes the rotational part $\overrightarrow{V_R} = \overrightarrow{v_1} - \overrightarrow{v_2} - a(\overrightarrow{\dot{\theta}_1} + \overrightarrow{\dot{\theta}_2}) \times \vec{n}$ where the subscripts 1 and 2 stand for the two colliding particles at a given time. Again, the normal restitution coefficient is obtained through $e = -|\vec{n} \cdot \overrightarrow{V'_R}|/|\vec{n} \cdot \overrightarrow{V_R}|$ and the tangential restitution coefficient by $\beta = -|\vec{n} \times \overrightarrow{V'_R}|/|\vec{n} \times \overrightarrow{V_R}|$. Another way to express the tangential restitution coefficient is to introduce the angle γ between $\vec{n}$ and $\overrightarrow{V_R}$ (**Figure 12**), we have for disks the relation [31]: $1 + \beta = -3(1 + e)\mu \cot(\gamma)$. The initial slope of β versus cot (γ) allows the computation of μ, the friction coefficient.

We have obtained experimentally an average value of $e = 0.64 \pm 0.03$. Although it is sometimes noticed in such situation [24, 32], we did not observe for the disks we used any clear dependence of e on the relative impact velocity. The experimental determination of the restitution coefficient α, between a particle and the walls of the cell reports a value $\alpha = 0.71 \pm 0.04$. We were also able to determine the behavior of the experimental tangential restitution coefficient as a function of $cot(\gamma)$.The results are presented in **Figure 13**. From the initial slope, one can compute an average value for the friction coefficient when particles are at contact: $\mu = 0.14 \pm 0.01$. As most of the binary collisions are head-on collisions (due to shape of the experimental cell), we have decided to take an average value of the tangential restitution $\beta = 0.7 \pm 0.05$.

The density and local velocity profiles of particles within the cell can be determined again from the positions of particles. In **Figure 14**, we have plotted the profiles of the x- and y- components of the disks' velocity. The area fraction of particles in the center of the cell is almost twice the initial one while close to the top and bottom walls, the value found is smaller. This is directly related to the inelastic nature of collisions which form clusters of particles [33, 34]. For this

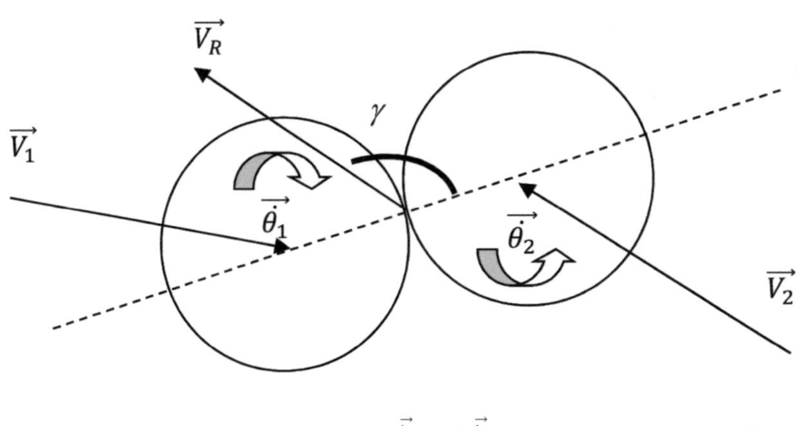

Figure 12. Sketch of a binary collision. $\overrightarrow{v_1}$ and $\overrightarrow{v_2}$, and $\overrightarrow{\dot{\theta}_1}$ and $\overrightarrow{\dot{\theta}_2}$ represent, respectively, the linear and rotational velocities of the particles before and after impact. $\overrightarrow{V_R}$ is the relative velocity and n the normal direction at collision. The impact angle γ is defined from n to $\overrightarrow{V_R}$.

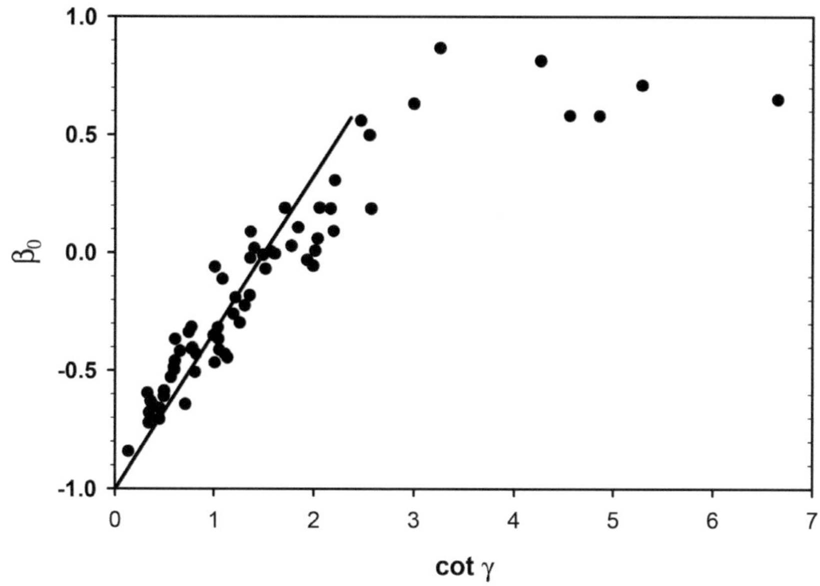

Figure 13. Experimental behavior of the tangential restitution β as a function of $cot(\gamma)$, where γ is the angle between the normal direction at contact and the direction of the relative velocity. The plain line is a linear regression used to compute the value of the friction coefficient when particles are at contact.

experimental study, the cell may again be divided into two different and well-identified regions [35]: a central one that we will refer as the "cold" area and the ones close to the top and bottom walls (where the energy is injected into the medium), referred to "hot" areas. In the following, subscripts H and C will be used, respectively, to identify the "hot" and "cold" regions of the cell. Considering all experiments, we have noticed first that the values found to characterize the "hot" zone by the height h_H were not related to the amplitude of vibration as one could expect, and second that an average value $h_H = 9$ mm ≈ 3.5a was acceptable in our experimental situations. Moreover, computation of the mean free path of particles in the "cold" zone gives a distance of about 12a (when using 12 disks) and 6a (for situations with 24 disks). These statements are found by considering the density of the "cold" region where typical values are found to be 13% (12 disks) and 30% (24 disks). We may conclude that the behavior of particles in the "cold" area is mainly governed by particle-particle collisions. We do not meet situations in which particles are moving through the bulk without being struck by another particle. Last, checking experimentally how the density profiles evolve with time does not present low-frequency oscillations like the situation reported in Ref. [36].

The temperatures of the granular medium are retrieved from the velocities of particles and from both contributions: the translation $T_{tr} = mV^2/2$, and the rotation $T_{rot} = I\omega^2/2$ (I the moment of inertia). In steady state, the temperature is calculated from a balance between two opposite fluxes: the energy brought to the medium by the particles in the "hot" areas of the cell and the dissipation in the bulk (i.e., the "cold" region). All experimental data presented below were retrieved from the "cold" region. To obtain reliable values, the whole set of the 20,000 images recorded are treated for each experimental value reported in this paper. We may also

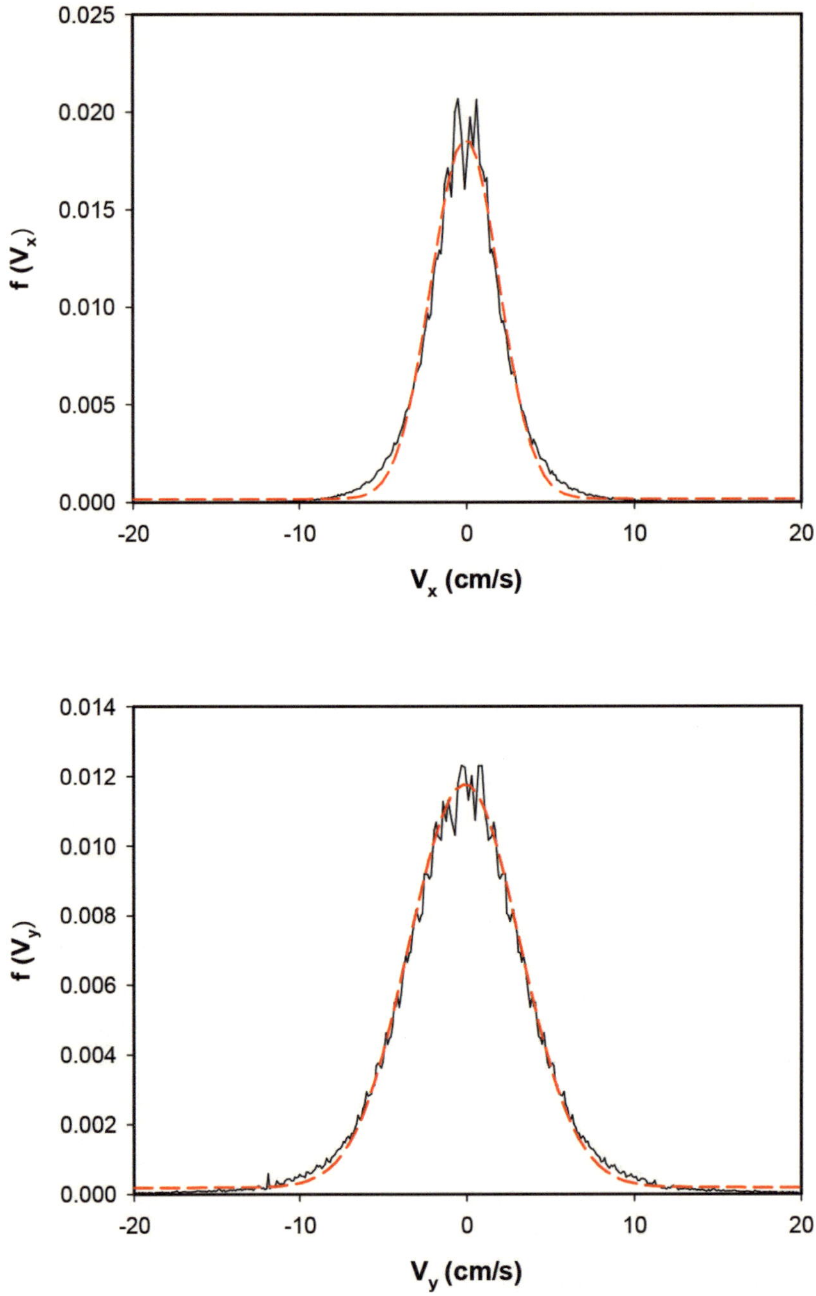

Figure 14. Velocity distributions of the component along the direction of vibration (y-direction) and transverse to it (x-direction). The experimental curves are drawn with plain lines. The dashed lines correspond to a Gaussian plot with the average velocity determined experimentally.

consider an undesired effect of g-jitter arising in parabolic flights but we will take it into account when comparing experimental results with theories. An average number N_H of particles present in the "hot" regions can be obtained directly from the density profiles at any time.

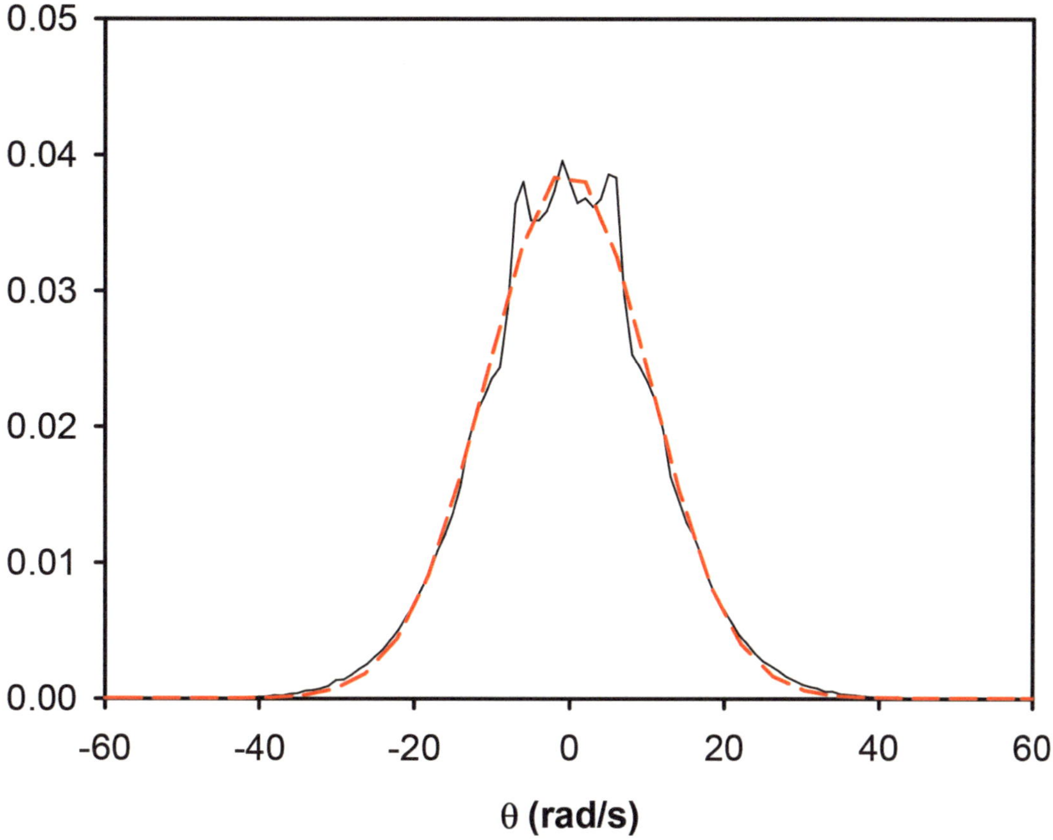

Figure 15. Typical angular velocity distribution of the particles (experiment: plain curve). The dashed line corresponds to the mathematical plotting of a Maxwell distribution, which includes the average angular velocity determined experimentally.

Typical experimental distributions for translation and rotational velocities are shown, respectively, in **Figures 14** and **15**: A Gaussian behavior can be observed. The dashed line on the figures represents the plots of the Gaussian distribution in which the experimental values of the squared velocities have been introduced, and one can notice the good agreement.

Due to the rectangular shape of the experimental cell used and to the relatively low area fraction, the main contribution to the temperature was expected to be found along the direction of the external vibration (the $y-$ direction). The temperature ratios T_y/T_x and T_{tr}/T_{rot} with $T_{tr} = (T_x + T_y)/2$, in terms of the maximum cell's velocity $A\omega$ ($\omega = 2\pi$n) for the two area fractions used, have been calculated. One may note that the ratio T_{tr}/T_{rot} is not drastically affected if one considers only T_y as the only contribution to the energy. For the fraction area of 16.6%, the ratio T_{tr}/T_{rot} is ranging from about 4 to 10, respectively, for maximum cell velocity from 20 to 40 cm/s, while T_y/T_x goes from 2 to 4 at maximum, while for the lowest fraction area (8.3%), T_{tr}/T_{rot} is ranging from 11 up to 24 and T_y/T_x from 5 to 7 only under the same conditions of maximum velocities. We shall analyze these experimental results by focusing first on the ratio T_y/T_x which is clearly dependent on the area fraction of the medium (the larger the ratio, the smaller the area fraction). Without any surprise, the temperatures obtained along the direction of the vibration are always larger than the ones in the transverse direction. This comes from the fact that the main part of energy injected is along the y-direction, and the relatively low area fraction does not allow the redistribution of this energy toward the perpendicular direction. At low area fraction, the particles can move easily and the y-direction drives the general motion. On the other hand, we also observe a net increase of the ratio T_y/T_x with the driving velocity of the cell, but less pronounced for the lower area fraction. However, the driving velocity is not the only parameter of the problem and the amplitude can also play a role. For example, the ratio $T_y/T_x = 1.98$ found was obtained for the smallest amplitude ($A = 0.556$ mm) and the largest frequency (60 Hz). Under the conditions of high frequency but small amplitude, we clearly see a concentration of particles in the central area of the cell, and consequently, the corresponding energy input is small. It then explains the low ratio found in such experiment, to be compared to a similar value of $A\omega = 0.22$ m/s but with much larger amplitude ($A = 2.3$ mm). We have set the frequency scale between 10 Hz and 30 Hz, and the corresponding amplitudes of vibration used are large enough to avoid the aggregation of particles in the center of the cell. Last, we can clearly identify the "cold" area straight from the density profiles. The second result is related to the ratio T_{tr}/T_{rot} which obviously increases with $A\omega$ and which is also dependent on the area fraction. The translational temperature is quite one order of magnitude larger than the rotational temperature. Because almost all collisions between particles are quite, head-on as reflected by the high value of T_y/T_x, can explain why the transfer from translational to rotational energy is rather weak, mainly at the lowest area fraction.

As a first step to describe the experimental behavior on granular temperatures, we can use existing theoretical models using a mean-field theory [37]. In this description, the rate of change of the temperature of a granular medium is determined through two coupled equations

$$\begin{cases} \dfrac{dT_{tr}}{dt} = J_{dr} + G[-AT_{tr}^{3/2} + BT_{tr}^{1/2}T_{rot}] \\ \dfrac{dT_{rot}}{dt} = 2G[B'T_{tr}^{3/2} - CT_{tr}^{1/2}T_{rot}] \end{cases} \tag{2}$$

where T_{tr} and T_{rot} represent, respectively, the translational and rotational temperatures, and $G = \frac{16}{\sigma\sqrt{\pi m}}\varphi g_2(\varphi)$ is related to the collision rate between particles; $g_2(\varphi)$ being the pair correlation function at contact. In two dimensions, $g_2(\varphi) = (1 - 7\varphi/16)/(1 - \varphi)2$. A, B, B', and C are constants, which depend only on the inelastic properties of the particles [38]. J_{dr} is the energy flux input into the medium and a homogeneous input of energy into the medium is assumed. These four constants are positive so that the minus signs are related to the loss of energy during the collision of particles. We state that the driving energy is acting on the translational temperature because of the preferred collisions with normal incidence. We may note that when the rotation mainly governs the behavior of the granular, J_{dr} is included in the second equation of (2) [39, 40].

Several inelastic modelizations were proposed by Herbst et al. [38] going from a simple consideration of a constant tangential restitution coefficient up to more complex ones where the tangential restitution depending on γ_{12} (the contact angle obtained neglecting the rotational velocities) or on the real contact angle γ. From the second equation of Eq. (2), the energy ratio T_{tr}/T_{rot} can be obtained considering the medium in steady state $dT_{rot}/dt = 0$, allowing to get the relation $T_{tr}/T_{rot} = C/B'$. In this equilibrium regime, $dT_{tr}/dt = 0$ and replacing T_{rot} in the first equation of Eq. (1) also give $T_{tr} = [CJ_{dr}/G(AC - BB')]^{2/3}$. Depending on the model used, the expressions of the constant C and B' can be evaluated and only depend on the inelastic properties of particles and to their inertia; however, the area fraction or the driving energy flux J_{dr} is never considered. Using our experimental values for the normal and tangential restitution and friction coefficient, we can numerically solve the models. Results show values of the ratio of T_{tr}/T_{rot} at maximum of 5, and more importantly, the results are not dependent on the maximum velocity of the cell. These models do not deal with an anisotropic temperature because their predictions are usually compared to numerical simulations where the energy is supposed to be added into the medium isotropically. This is why these models cannot represent our results.

When an external vibration is acting on the granular, it can be viewed as a medium which dissipates energy while energy is added into it through the vibration per unit time. The equilibrium temperature (and state) can be found from the equilibrium equation $J_{dr} + Q_d = 0$, where J_{dr} is the energy flux injected in the medium by the collisions of particles with the walls of the cell and Q_d, the energy flux dissipated during the binary collisions between particles. J_{dr} is found to act in the "hot" areas of the cell, while Q_d is computed in the bulk. The results obtained from experiment and geometry clearly show that the main energy input on particles occurs along the direction of the vibration. From our observations, we have considered the areas of energy input by defining two layers of thickness h_H and having the same width L_x. Then, the particles' density is much smaller than the one of the medium, and we have introduced N_H as the average number of particles present at any time. Thus, the bulk of the medium (i.e., the "cold" zone) reduces to dimensions $h_C = L_y - 2h_H$ where only $N_C = N - 2N_H$ particles are present at any time; the surface of this zone is then $S_C = h_C L_x$. In the "cold" zone, the dissipated energy depends on the collision frequency $f_E(T)$ which in turns depends on the temperature T of the medium $T = m\langle v_x^2 + v_y^2\rangle/2$. If we neglect the loss of energy coming from tangential restitution coefficient, the energy dissipated per collision is given by:

$$\Delta E_{pp} = m\frac{(e2-1)}{4}\langle[(\vec{v_1} - \vec{v_2})\cdot\vec{n}]^2\rangle = \frac{(e^2-1)}{2}T \qquad (3)$$

The frequency collision which is the inverse of the Enskog time is given in 2D by $f_E = \sqrt{2\pi}\frac{N_C}{S_C}\sigma g_2(\varphi)\langle v\rangle = \frac{2}{N_C}f_E^N$ where N_C/S_C represents the number density in the "cold" region and f_E^N is the number of collisions between N particles per unit time. Finally, the dissipated energy flux will be expressed as follows:

$$\begin{aligned} Q_d = f_E^N \Delta E_{pp} &= \frac{N_C^2}{h_C L_x}\frac{1-e^2}{2}\sigma g_2(\varphi)\sqrt{\frac{\pi}{m}}T^{3/2} \\ &= \frac{N_C^2}{h_C L_x}\frac{1-e^2}{4}\sigma g_2(\varphi)\sqrt{\frac{\pi}{m}}T_y^{3/2}\left(1+\frac{T_x}{T_y}\right)^{3/2} \end{aligned} \qquad (4)$$

To express the energy input into the medium, we must now take into account the flux coming from the collisions between the particles and the walls of the cell. When a collision occurs, the kinetic energy change for one particle is: $\Delta E_{pw} = m(v\prime_y^2 - v_y^2)/2$, where $v'_y 2$ and v_y^2, respectively, are the velocities of the particle after and before collision with the cell's wall. The cell is assumed to move with a velocity V_{dr}. The relative velocity equation gives $v' - V_{dr} = \alpha(V_{dr} - v)$, where α is the normal restitution coefficient between the particle and the wall. The change in kinetic energy of one particle may then be rewritten as $\Delta E(v_y, V_{dr}) = \frac{m}{2}[(1+\alpha)^2 V_{dr}^2 - 2(1+\alpha)V_{dr}v_y - v_y^2(1-\alpha 2)]$ and the energy flux j_{dr}, associated with particles going toward the wall, can be expressed as $j_{dr} = \frac{N_H}{2h_H} v_y \Delta E(v_y, V_{dr})$, where we have assumed that $N_H/2$ particles are going toward the wall. The net energy flux for a given wall velocity is then obtained by averaging the flux of the incoming particles with the velocity distribution function, $f(v_y)$ associated with the "cold" region and integrating on the velocities directed towards the wall:

$$J_{dr}(V_{dr}) = \int_0^\infty j_{dr} f(v_y) dv_y \tag{5}$$

where the distribution function of the velocity is the Gaussian one retrieved from experiments (**Figure 14**).

The integral (5) over the velocities gives the following result

$$J_{dr}(V_{dr}) = \frac{m}{4}\frac{N_H}{h_H}[(1+\alpha)^2 V_{dr}^2 I_1 - 2(1+\alpha)V_{dr} I_2 - (1-\alpha 2)I_3] \tag{6}$$

I_1, I_2, and I_3 are the integrals $\int_0^\infty v_y^i f(v_y)\, dv_y$ $(i = 1..3)$ which are, respectively, given by:

$$I_1 = \sqrt{\frac{T_y}{2\pi m}} I_2 = \frac{T_y}{2m} I_3 = \left(\frac{T_y}{m}\right)^{\frac{3}{2}} \sqrt{\frac{2}{\pi}} \tag{7}$$

We may assume that the particles go from the bulk towards the "hot" areas (double collisions are neglected) so that with the average velocity found experimentally and that we are using in the comparisons. The last thing to do is to compute the average on the wall velocity: the linear term in V_{dr} cancels while the term related to V_{dr}^2 averages to $(A\omega)2/2$. Multiplying by 2 (because of 2 moving walls) gives the following expression for the injected flux of energy:

$$J_{dr} = m\frac{N_H}{2h_H}\left[(1+\alpha)2(V_{dr})2\sqrt{\frac{T_y}{2\pi m}} - (1-\alpha 2)\left(\frac{T_y}{m}\right)^{\frac{3}{2}}\sqrt{\frac{2}{\pi}}\right] \tag{8}$$

For perfectly elastic walls ($\alpha = 1$), the expression proposed by Soto [35] for a sinusoidal vibration taking for their function $q\left(T/m(A\omega)2\right)$ is recovered; the constant $q = \sqrt{2/\pi} = 0.8$ is a very good approximation for our experimental conditions and our values of $T/m(A\omega)2$. The additional contribution to the energy input from g-jitter (even if it has quite no impact on the velocities of the free floating particles in microgravity) can create an additional contribution to the wall velocity and so to the energy injected into the medium. This can be estimated as $\langle \delta V^2 = 0.005\, m^2/s^2 \rangle$ [41]. Since it does not exceed 15% of V_{dr}^2 in the worst case, it was neglected. The equilibrium between injection (Eq. (8)) and dissipation (Eq. (4)) gives:

$$T_y = \frac{\frac{N_H}{2h_H}(1+\alpha)2}{\frac{N_C^2}{h_C L_x}\pi\sigma g_2(\varphi)\frac{1-e2}{2}\left(1+\frac{1}{R_T}\right)^{3/2}+2\frac{N_H}{h_H}(1-\alpha 2)} m(A\omega)2 \quad (9)$$

The temperature is proportional to the square of the amplitude of the driving velocity as it should have been shown in Ref. [13]. The densities in the "cold" and "hot" areas are known so that we can compare the predictions of Eq. (9) with our experimental values of T_y calculated. To consider the dissipation due to the tangential restitution coefficient β, we introduce an effective restitution coefficient r_e proposed by McNamara and Luding [41] instead of e in Eq. (9): $r_e = \sqrt{e^2 - \left(q(1-\beta^2)\right)/(1+2q-\beta)}$. Using $q = 0.5$ for a disk and $\beta = 0.7$;, we obtain $r_e = 0.462$ instead of $e = 0.64$. The comparison between the theoretical temperatures T_y obtained from Eq. (9) with the experimental ones calculated in the "cold" region is presented in **Figure 16**.

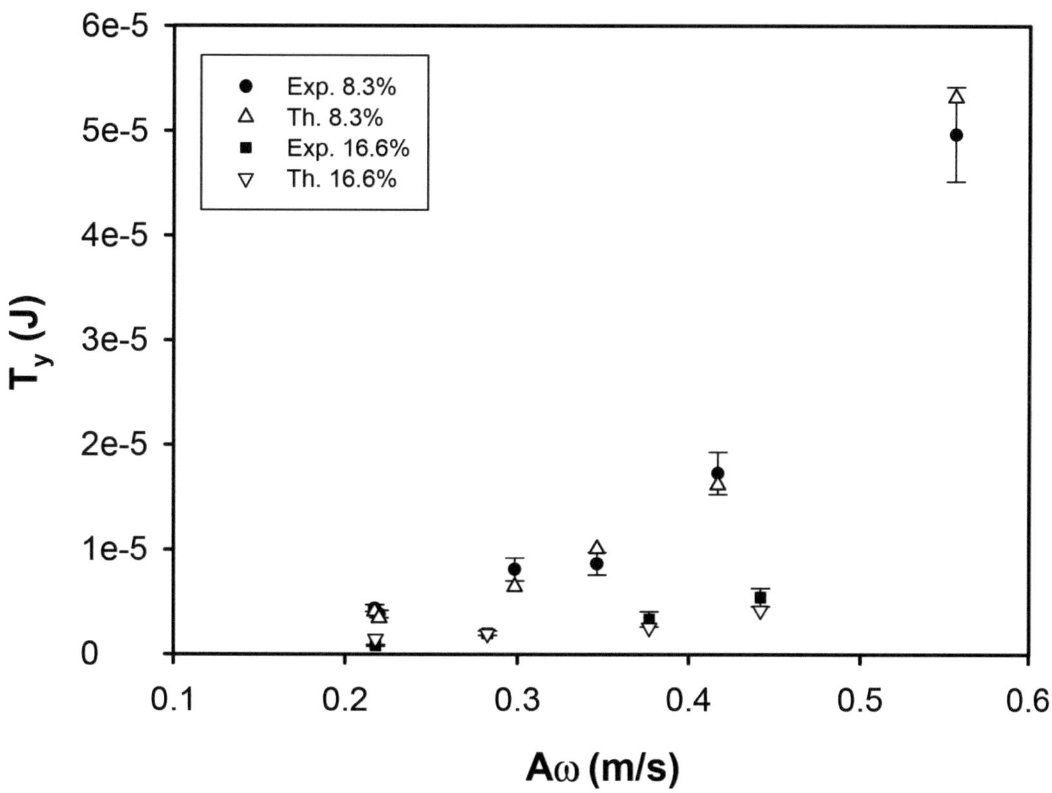

Figure 16. Comparison of the equilibrium temperature computed from Eq. (9) as a function of the driving velocity of the cell ($A\omega$).

With the two area fractions, we have used in this study, one can see that the agreement is good. Of course, to improve the theoretical prediction of the temperature with the driving velocity, being able to predict the density N_H/h_H close to the top and bottom walls instead of considering the experimental value obtained from the profiles.

The anisotropy found for the temperatures created by a vibrating wall is scarcely reviewed in the literature. A recent experiment with a setup including a 3D-cylindrical [42] where the anisotropic behavior of the ratio $R_T = T_y/T_x$ is reported as a function of the volume fraction of particles shows a strong increase for values below 10%, but it still remains smaller than our results. Moreover, a theoretical study including two different Maxwellian distributions for parallel and perpendicular directions about the vibration and a density along the vibration axis is presented in Ref. [43]. A balance between energy fluxes along the direction of vibration and perpendicular to it gives the ratio R_T and predicts that, for perfectly elastic walls, this ratio would only depend on the restitution coefficient. This is not our experimental situation where the ratio R_T is much larger for the lowest density. Thus, it impossible to relate this theoretical approach with our study since our density profile is much different for a one driven by gravity. Nevertheless, we may predict values of R_T if we assume a constant density in the "cold" area.

4. Conclusion

We have reported experimental investigations on the dynamics of a model granular medium. Experiments have been performed in a low-gravity environment. The cell containing the medium is subjected to external vibration which drives the collective motion of the particles. As the dynamical behavior of the medium is driven by the kinematics of the particles, high-speed video recording coupled to an individual particle tracking technique allows to obtain the trajectory of each particle. From these raw data, the inelastic parameters of the particles which are at the origin of the dynamics of the whole medium can be retrieved as well as a direct measurement of the energy (or temperatures). We have found that depending on the type of particles used, the normal restitution coefficient can be dependent on the relative impact velocity between two particles but not always. One way to characterize the inelastic nature of the collisions is to look to the energy decay once the medium is freely evolving. We have obtained smaller experimental relaxation times of this energy than the ones predicted by theories at least if we do not take into account the velocity dependent of this restitution coefficient. It is also interesting to note that the effect played by the rotation of the particles can significantly affect the whole behavior of the medium. In particular, we have reported the translational temperatures along and perpendicular to the direction of vibration as well as the rotational temperatures. When compared to existing theories, it appears that there are significant differences which also depend on the driving velocity and on the concentration of the medium. Two major points on the comparison can be raised: First the density is not homogeneous in the cell and second the translational velocities are much higher in the direction of vibration than perpendicular to it (versus a homogeneous input of energy as considered in theories). We report that the balance of the energy fluxes along the vibration can correctly represent the behavior of the granular temperature with the driving velocity of the cell and with the area fraction. In this balance, the contribution of the tangential velocities to the dissipation must be considered. At least the distinction between the dissipation due to the collisions between the particles which is proportional to the average temperature $T = (T_x + T_y)/2$ and the driving flux, which depends only on T_y, was introduced, but on the basis of the experimental ratio T_y/T_x. This ratio increases when the volume fraction decreases and it also depends on the driving velocity. A theoretical determination of T_y/T_y which could reproduce these behaviors should involve the non-elastic collisions with the lateral walls, but is let for a future developments.

Acknowledgements

The authors like to thank NOVESPACE and the CNES for giving them the possibility to board the A300-zero G in order to perform the experimental study.

Author details

Yan Grasselli[2*], Georges Bossis[1], Alain Meunier[1] and Olga Volkova[1]

*Address all correspondence to: yan.grasselli@skema.edu

1 Laboratoire de la Matière Condensée, University of Nice Sophia Antipolis, Parc Valrose, France

2 SKEMA Business School, University of Côte d'Azur, Sophia-Antipolis, France

References

[1] Pöschel T, Luding S, editors. Granular Gases. Springer; Berlin, 2001

[2] Pöschel T, Brilliantov NV, editors. Granular Gas Dynamics. Springer; Berlin, 2003

[3] Barrat A, Trizac E, Ernst MH. Granular gases: Dynamics and collective effects. Journal of Physics: Condensed Matter. 2005;**17**:S2429

[4] Aranson IS, Tsimring LS. Patterns and collective behavior in granular media: Theoretical concepts. Reviews of Modern Physics. 2006;**78**:641

[5] Puglisi A, editor. Transport and Fluctuations in Granular Fluids: From Boltzmann Equation to Hydrodynamics Diffusion and Motor Effects. Springer; Berlin, 2014

[6] Kudrolli A, Wolpert M, Gollub JP. Cluster formation due to collisions in granular material. Physical Review Letters. 1997;**78**:1383

[7] Opsomer E, Ludewig F, Vandewalle N. Phase transitions in vibrated granular systems in microgravity. Physical Review E. 2011;**84**:051306

[8] Falcon E, Wunenburger R, Evesque P, Fauve S, Chabot C, Garrabos Y, Beysens D. Cluster formation in a granular medium fluidized by vibrations in low gravity. Physical Review Letters. 1999;**83**:440

[9] Falcon E, Aumaıtre S, Evesque P, Palencia F, Lecoutre-Chabot C, Fauve S, Beysens D, Garrabos Y. Collision statistics in a dilute granular gas fluidized by vibrations in low gravity. Europhysics Letters. 2006;**74**:830

[10] Noirhomme M, Opsomer E, Vandewalle N, Ludewig F. Granular transport in driven granular gas. The European Physical Journal E. 2015;**38**:9

[11] Opsomer E, Noirhomme M, Vandewalle N, Falcon E, Merminod S. Segregation and pattern formation in dilute granular media under microgravity conditions. Nature. Article number: 1, 2017;**3**

[12] Bossis G, Grasselli Y, Volkova O. Granular rheology in zero gravity. Journal of Physics: Condensed Matter. 2004;**16**:3279

[13] Bhateja A, Sharma I, Singh JK. Scaling of granular temperature in vibro-fluidized grains. Physics of Fluids. 2016;**28**:043301

[14] Pathak SN, Jabeen Z, Das D, Rajesh R. Energy decay in three-dimensional freely cooling granular gas. Physical Review Letters. 2014;**112**:038001

[15] Huthmann M, Aspelmeier T, Zippelius A. Granular cooling of hard needles. Physical Review E. 1999;**60**:654

[16] Villemot F, Talbot J. Homogeneous cooling of hard ellipsoids. Granular Matter. 2012;**14**:91

[17] Harth K, Kornek U, Trittel T, Strachauer U, Home S, Will K, Stannarius R. Granular gases of rod-shaped grains in microgravity. Physical Review Letters. 2013;**110**:144102

[18] Yan-Pei C, Evesque P, Mei-Ying Chin H. Breakdown of energy equipartition in vibro-fluidized granular media in micro-gravity. Physics Letters. 2012;**29**:074501

[19] Brilliantov NV, Pöschel. Kinetic Theory of Granular Gases. Oxford University Press; 2004

[20] Tatsumi S, Murayma Y, Hayakawa H, Sano M. Experimental study on the kinetics of granular gases under microgravity. Journal of Fluid Mechanics. 2009;**641**:521

[21] Grasselli Y, Bossis G. Three-dimensional particle tracking for the characterization of micrometer-size colloidal particles. Journal of Colloid and Interface Science. 1995;**1**:269

[22] Gondret P, Lance M, Petit L. Bouncing motion of spherical particles in fluids. Physics of Fluids. 2002;**14**:268

[23] KantakAdvait Ashok. Wet particles collisions [thesis]. University of Colorado at Boulder: 2005. DOI: AAT 3190381

[24] Sorace CM, Louge MY, Crozier MD, Law VHC. High apparent adhesion energy in the breakdown of normal restitution for binary impacts of small spheres at low speed. Mechanics Research Communications. 2009;**36**:364

[25] Chapman S, Cowling TG, editors. The Mathematical Theory of Nonuniform Gases. London: Cambridge University Press; 1960

[26] Miller S, Luding S. Cluster growth in two-and three-dimensional granular gases. Physical Review E. 2004;**69**:031305

[27] Hou M, Liu R, Zhai G, Sun Z, Lu K, Garrabos Y, Evesque P. Velocity distribution of vibration-driven granular gas in Knudsen regime in microgravity. Microgravity Science and Technology. 2008;**20**:73

[28] Tatsumi S, Murayama Y, Sano M. Experimental Study of the Freely Evolving Granular Gas under Microgravity Condition. AIP Conference Proceedings. 2008;**1027**:923

[29] Maaß C, Isert N, Maret G, Aegerter CM. Experimental investigation of the freely cooling granular gas. Physical Reviews Letters. 2008;**100**:248001

[30] Chen Y, Hou M, Evesque P, Jiang Y, Liu M. Asymmetric velocity distribution in boundary-heating granular gas and a hydrodynamic description. Powders & Grains. 2013;**1542**:791

[31] Labous L, Rosato AD, Dave RN. Measurements of collisional properties of spheres using high-speed video analysis. Physical Review E. 1997;**56**:5717

[32] Grasselli Y, Bossis G, Goutallier G. Velocity-dependent restitution coefficient and granular cooling in microgravity. European Physical Letters. 2009;**86**:60007

[33] Das S, Puri S. Pattern formation in the inhomogeneous cooling state of granular fluids. Europhysics Letters. 2003;**61**:749

[34] Evesque P. Powders & Grains. 2001;**12**:60

[35] Soto R. Granular systems on a vibrating wall: The kinetic boundary condition. Physical Reviews E. 2004;**69**:061305

[36] Rivas N, Luding S, Thornton AR. Low-frequency oscillations in narrow vibrated granular systems. New Journal of Physics. 2013;**15**:113043

[37] Jenkins T, Zhang C. Kinetic theory for identical, frictional, nearly elastic spheres. Physics of Fluids. 2002;**14**:1228

[38] Herbst O, Cafiero R, Zippelius A, Herrmann HJ, Luding S. A driven two-dimensional granular gas with Coulomb friction. Physics of Fluids. 2005;**17**:107102

[39] Falcon E, Bacri J-C, Laroche C. Equation of state of a granular gas homogeneously driven by particle rotations. Powders & Grains. 2013;**1542**:815

[40] vanNoije TPC, Ernst MH. Velocity distributions in homogeneous granular fluids: The free and the heated case. Granular Matter. 1998;**1**:57

[41] McNamara S, Luding S. Energy flows in vibrated granular media. Physical Review E. 1998;**58**:813

[42] Windows-Yule CRK, Parker DJ. Boltzmann statistics in a three-dimensional vibrofluidized granular bed: Idealizing the experimental system. Physical Review E. 2013;**87**:022211

[43] van der Meer D, Reimann P. Temperature anisotropy in a driven granular gas. Europhysics Letters. 2006;**74**:384

Chapter 5

Particle Jetting Induced by the Impulsive Loadings

Kun Xue, Xiaoliang Shi, Kaiyuan Du and Haoran Cui

Additional information is available at the end of the chapter

http://dx.doi.org/10.5772/intechopen.68795

Abstract

Particle rings/shells/cylinders dispersed by the radial impulsive loadings ranging from strong blast waves to moderate shock waves form a dual coherent jetting structure consisting of particle jets which have different dimensions. In both circumstances, the primary jets are found to initiate from the inner surface of particle layers and propagate through the thickness of particle layers, which are superimposed by a large number of much smaller secondary jets initiating from the outer surface of particle layers upon the reflection of the shock wave. This chapter first presents a summary of the experimental observations of the hierarchical particle jetting mainly via the cinematographic techniques, focusing on the characteristics of the primary particle jet structure. Due to the distinct behaviors of particles subjected to the strong blast and moderate shock waves, specifically solid-like and fluid-like responses, respectively, the explosive and shock-induced particle jetting should be attributed to distinct mechanisms. A dual particle jetting model from the perspective of continuum is proposed to account for the explosive-induced particle jetting. By contrast the shock-induced particle jetting arises from the localized particle shear flows around the inner surface of particle layers which result from the heterogeneous network of force chains.

Keywords: particle jetting, blast wave, shock wave, force chains, discrete element method, multiphase flows

1. Introduction

When particles are dispersed by an impulsive pressure loading, the expanding particle cloud typically forms a nonuniform structure that takes the form of particle jets whose leading edges are agglomerates of constituent grains [1–12]. A host of experimental evidence from a wide range of sources shows that the expanding cloud of explosively disseminated material comprises of "particles" or fragments that have significantly different dimensions from those associated with the original material as shown in **Figure 1(a)** and **(b)** [1–7, 10–13]. Photographic evidence shows characteristic jets or fingers behind these expanding fragments. These coherent conical particle jets travel ballistically as shedding mass along the trajectories with increasingly diffuse edges.

Particle jetting has been widely observed in volcanic eruptions, supernovae, explosion of landmines, thermobaric explosion (TBX), fuel-are explosion (FAE), and dense inert metal explosive (DIME) [14–17]. The formation of particle jetting has also been observed during the impact of solid projectile on granular media [18]. The structure of particle jets in terms of the jet number of size is important to the viability of many applications. For instance, the strength of TBX and FAE needs to be enhanced by the after-burning of the reactive particles dispersed in the payload cloud. The detonation performance of the particle laden cloud depends on both the shape and concentration of the cloud which in turn is a result of the jet mixing [19]. In contrast with the large-scale injury radius of TBX and FAE, DIME utilizes the high-speed dense metal–particle jets to hit the targets in close range. Thus the momentum of particle jets determines the lethal radius. Another opposite application is mitigation of the blast pressure

Figure 1. Explosive dispersal of dry (a) and wetted glass beads (b) using cylindrically stratified configurations [12]. Shock dispersal of flour (c) and quartz sand (d) particles using semi-two-dimensional configurations.

(both prompt and quasi-static) associated with the detonation, since a commonly used technique to reduce effects of blast from explosives is to surround the explosive with a layer of liquid, powder, or a slurry mixture of the two. Drag is seen as a potential mechanism to transfer energy from the blast wave to the disseminated particles or droplets so the size of particles or formation of jets is important in determining the efficiency of this mechanism. Frost and Zhang have reviewed many of the processes occurring in heterogeneous blast including jet formation [15, 17, 20, 21].

Extensive experimental investigations of the explosive- or shock-induced particle jetting mainly using cinematographic techniques provide fundamental data regarding both structure and evolution of particle jets. Closer look into the high-speed photos of either explosive- or shock-induced particle jetting reveals a dual structure (see **Figure 1(c)**) [1, 2, 12]. Primary jets initiated on the inner surface of the particle layers take shape during the first dozens of microseconds after the detonation of the central explosive evidenced by the light stripes detected from the radiographs of the explosive dispersal of particle shells [5, 6]. Upon the reflection of the shock wave on the outer surface of particle layers, a large number of smaller jets begin to burgeon from the outer surface and quickly develop into a full bloom [3]. The dominant primary jets are expelled from the outer surface and overtake the smaller secondary jets, merging of secondary jets occurring through the aerodynamic interaction. The respective evolutions of the primary and secondary jets are not so distinguishable from the radiographs and high-speed photos of the explosive dispersal of particles (see **Figure 2**). But the statistic distribution of jet size unravels two distinctive peaks representing the primary and secondary jets, respectively [13]. In order to overcome the difficulties in distinguishing the primary and secondary jets, a semi-two-dimensional configuration based on the Hele-Shaw cell that will be discussed in Section 3.1 was employed to access the evolution of both sets of jets subjected to the radial shock loading. Although the overpressure of weak shock waves is several orders of magnitude lower than that of blast waves, the formation and evolution of the primary and secondary jets as shown in **Figure 3** have astonishingly similar characteristics in terms of the initiation sequence and the signature structure [2]. Whereas whether or not the jetting process in these two extreme conditions follow the same path is still debatable.

Great efforts have been devoted to investigate the dependence of the jet number on a variety of parameters, including the configuration of charge, the mass ratio of the payload and the explosive (M/C), the inner and outer radius of particle layers, the particle material and size, and the moisture content, etc., mainly in the case of explosive dispersal of particles [2, 4, 6, 8, 13]. Specifically, Zhang et al. found that the numbers of primary and secondary jets dispersed by the 44 mm diameter of central explosive cylinder are 1.8 and 1.5 times those with the 10 mm diameter of explosive [14]. Frost et al. found that the jetting phenomenon is much more visible in cases of explosive dispersal of brittle or ductile powders, such as quartz sand, glass beads, SiC powders, aluminum powders, copper powders, compared with rigid and hard powders, like stain steel particles that are dispersed into the particle cloud rather than particle jets [10]. Frost and Xue both found that the addition of the interstitial water/oil significantly increases the jet number [12, 13].

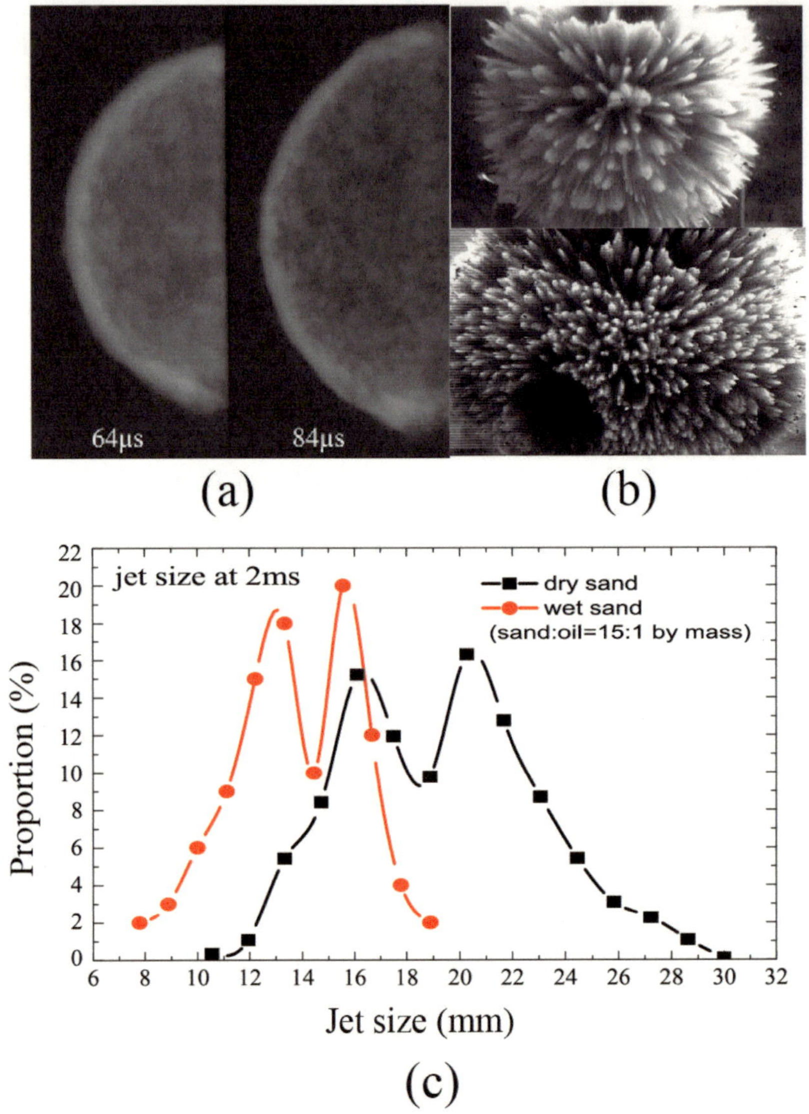

Figure 2. Radiographs and high-speed photos of explosive dispersal of glass beads (a) [5], dry (above panel of (b)) and wet (bottom panel of (b)) quartz sand grains [13]. (c): the statistic distribution of size of the explosive induced dry and wet sand jets at t = 2ms [13].

Some fundamental problems need to be addressed in this regard. First, several variables are correlated rather than independent so that it is impossible to single out the effect of the individual variable. For instance, changing the inner or outer radius of the particle layers or particle materials would inevitably alter the M/C that proves to be key factor determining the jet number. Rodriguez et al. proposed an alternative way to measure the effects of pertinent factors [2]. Acceleration of the outer surface of particle ring, which is a function of a variety of parameters, is found to be key determining factor. Therefore, choosing some proper dynamic

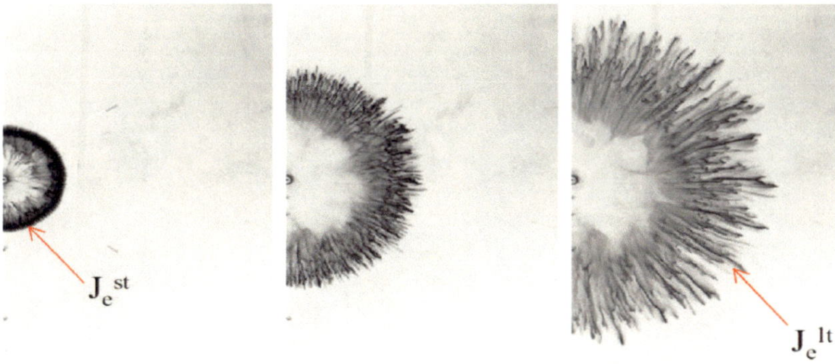

Figure 3. High-speed photos of semi-two-dimensional shock-induced particle jetting [2].

variables instead of structural parameters may well provide a new perspective in this regard, but entailing a thorough understanding of the physics underlying the particle jetting.

Second, distinguishing the primary and secondary jets from the radiographs or high-speed photos of the explosive dispersal of particles using either spherical or cylindrical stratified configurations is so difficult if not impossible that the validity of the experimental results is questionable thanks to the superimposition of two sets of jets on the timescale of microseconds. Adopting a semi-two-dimensional configuration in which particle rings are dispersed by the radial propagating shock waves seems to be promising approach to this problem. Besides, no detonation product gases obscuring the particle jets and substantially prolonged duration of jets facilitate the observation of particle jetting. But to what extent the shock-induced particle jetting can mimic that driven by the central explosion is quite questionable taking into account that the overpressure of shock waves is several orders lower than that of blast waves.

Predicting the jet number entails the knowledge of the mechanisms governing the primary and secondary jets, respectively. Several theories have been put forward, but understanding the origin of particle jetting still remains a significant challenge [3, 8, 9, 13, 22, 23]. The timescale for the formation of primary jets predicted by the Rayleigh-Taylor instability is much slower than the experimental observation [6]. Another interface instability theory involves the perturbation on the inner and outer surfaces of the particle layers that act as the microjets precipitating the macrojets propagating into the bulk. Riple et al. demonstrated the evolution of the initial perturbation (see **Figure 4**) into well-developed jets and argued that the casing fragments and other imperfections may provide the initial perturbation [3]. However, particle jetting occurs regardless of the presence of the inner and outer casings and shows similar structure. Certain intrinsic imperfections with the length scale similar to the jets should exist if this theory holds. An increasing number of investigators have focused their attention on the bulk fracture of powder bed. Frost et al. postulated that the breakup of a layer of particles at high strain rates was governed by a balance of expansion inertia effects tending to fracture the layer versus viscous dissipation that tends to maintain the stability of the layer [24]. Along this line, Xue et al. developed a theoretical model account for the instability onset of the expanding powder shell [13]. Milne et al. conjectured that the powder is explosively compacted into

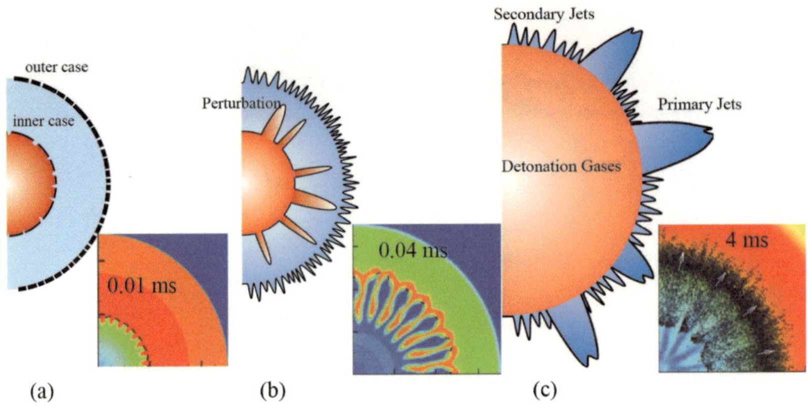

Figure 4. Schematics of the formation of primary and secondary jets caused by the inner and outer cases, respectively. (a): fragmentation of the inner and outer cases; (b): formation of primary and secondary jets arising from the gaps between fragments of inner and outer cases, respectively; (c): primary jets overtake the secondary jets. Inset: the corresponding snapshots from the hydrodynamic simulations of Ripley et al. [3,19].

a brittle solid which then forms cracks as the shell expands [5]. This conjecture is consistent with the observations that the primary jetting occurs during the first wave transit times. The major obstacle of this argument is that the compacted powder cannot sustain the tension or the surface energy, both among the essential components comprising the brittle fragmentation of solids. A few attempts try to understand the secondary jets, and the earlier works of Ripley et al. focused on the Richtmyer-Meshkov instability (RMI), which showed well-defined persistent jetting structures matching the number of prescribed outer surface perturbations [3]. However, the timescale for formation was slow and the surface instability did not propagate into the bulk [3]. Xue et al. modified the hollow sphere expansion model that originally accounts for the spallation of shocked solids so that the external particle jetting can be seen as parallel to the solid spallation [22].

Despite the resembling phenomenal features sheared by the explosive- and shock-induced particle jetting, the shock interaction with particles in the explosive dispersal is substantially stronger than that in the weak shock dispersal. In the former case, particles are compressed into solids with the density almost same as that of the constituent materials when the particle jetting commences. It suggests that a continuum approach is appropriate to model the explosion-driven particle jetting. By contrast, the weak shock wave only initiates the homogeneous or localized unsteady flows on the particle scale. The shocked particles behave more like fluids rather than solids. Unsteady and heterogeneous particle flows occurring during the weak shock interaction with particles entail a particle scale approach. Xue et al. described the particle scale formation and evolution of particle jets via the discrete element method (DEM), shedding some lights on the distinctive origins of the shock-induced particle jetting [25].

This chapter first reviews the up-to-date understanding of the phenomenology and physics of the particle jetting in both explosion-driven and shock-induced cases. Special attention is focused on theoretical progresses in unraveling the mechanism behind the respective

http://dx.doi.org/10.5772/intechopen.68795

particle jetting and establishing models account for the onset of jetting, which is elaborated in Sections 2 and 3. Further work and possible breakthrough in this regard would be discussed in Section 4. The conclusion is presented in Section 5.

2. Explosion-driven particle jetting

2.1. Strong shock interaction with particles

One generally accepted fact of the explosive-driven particle jetting is that particle instabilities occur during the first dozens of the microseconds after the detonation of the central explosive. It is thus necessary to elucidate the interactions between particles, shock waves, and detonation product gases. Hydrodynamic simulations [22] have been performed to reveal the evolution of dry and saturated sand layers surrounding the spherical central explosive (TNT or HXM), the configuration illustrated in **Figure 5(a)**. In order to accurately describe the dynamic responses of wet sand with different degrees of saturation β, we adopted a modified version of Laine and Sandvik model developed by Grujicic et al. [26] to account for the effect of moisture content via explicitly incorporating the degree of saturation in the equation of state (EOS) and the strength model. Given the relative incompressibility of the water phase, the compressibility of the wet sand is increasingly reduced with the degree of saturation as illustrated by the EOS of the wet particles with varying saturation (see **Figure 5(b)**). Besides, the wet sand's yield stress is reduced due to the moisture-induced interparticle lubrication effects leading to a reduced effective friction coefficient (see **Figure 5(c)**. For details of the modified compaction model, readers can be referred to Refs. [26, 27] (see **Figure 6**).

The evolvement of the sand shell upon the blast wave can be well embodied by the variations in its radial density profile as shown in **Figure 3**. The sequence of events basically resembles those occurring in the shock-loaded water shell described by Milne et al. [5, 6]. When the shock front reflects upon the outer surface of the particle shell, the rarefaction wave travels back into particles and pulls away a thin spall layer moving forward into air. The compressive

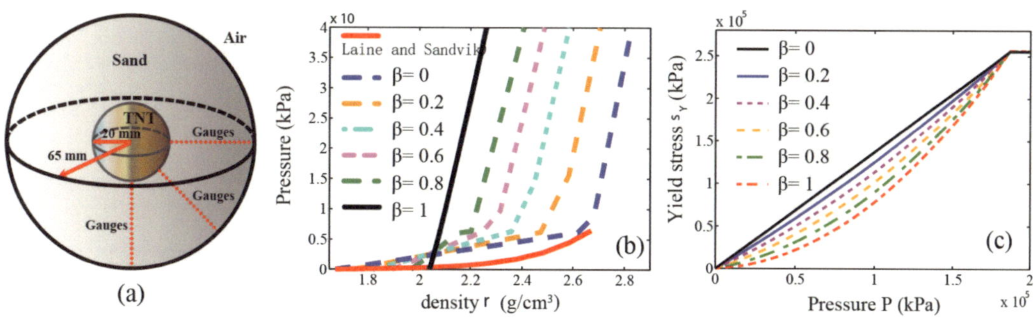

Figure 5. (a) Schematic of the spherical stratified configuration used in the hydrodynamic simulations. (b) EOS curves of the sand with varying degree of saturation. (c) Variations in dependence of the sand's yield stress on the pressure with increasing moisture contents [22].

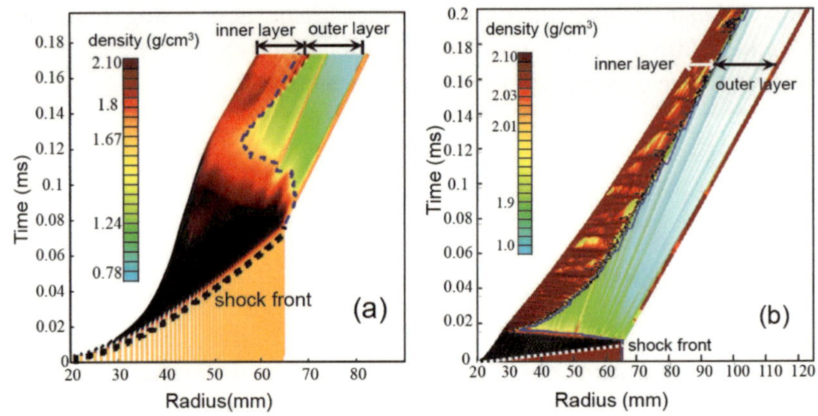

Figure 6. Evolutions of density profiles in dry sand (a) and saturated sand (b) after the detonation of the central explosive (TNT) [22].

stresses in the compacted particles are relaxed in the wake of the rarefaction wave accompanied by the rapid decrease of the packing density. The expansion of detonation product gases sends a shock wave into the particles, which arrests the rarefaction wave in its path in the case of dry sand or recompact particles diluted by rarefaction wave in the case of saturated sand. As a result, besides the outmost thin spall layer, the particle shell evolves into two distinct layers, namely the inner compact layer and outer dilute layers. The inner compact layer retains the maximum density almost as that of pure quartz and expands as an incompressible shell during a relatively long time, at least during the first hundred of microseconds after the detonation of central explosive. The hypothesis is supported by the consistent velocity across the thickness of the inner compact layer (see **Figure 7(a)** and **(b)**). Opposedly, particles inside the outer dilute layer lose the persistent contacts in the wake of the rarefaction wave. The

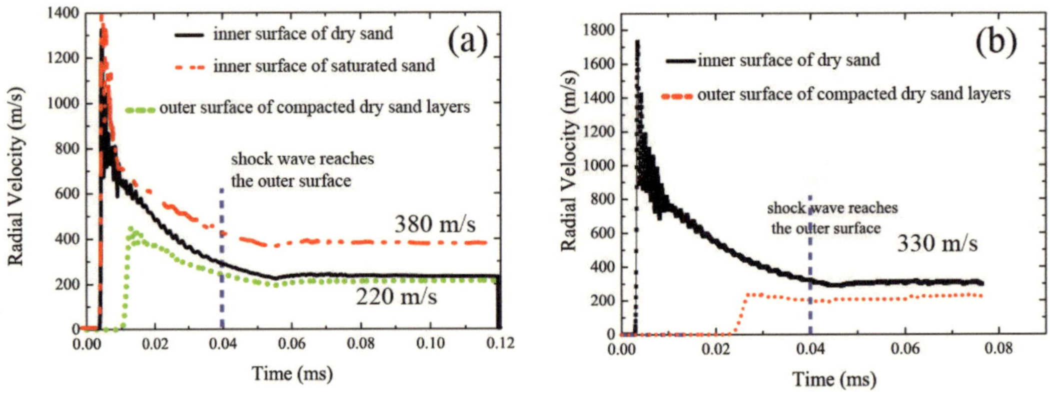

Figure 7. Evolutions of velocities of the inner and outer surfaces of the compact dry/saturated sand layer driven by the detonation of central TNT (a) and HXM (b).

http://dx.doi.org/10.5772/intechopen.68795

mass ratio of the compact and dilute layers depends on the geometry and the composition of granular shell, as well as the strength of central explosive.

Due to the trivial compressibility of saturated sand, the acceleration of the compact saturated sand layer is much stronger than that of dry sand since less shock energy is dissipated among the compaction. The expanding velocity of the compact saturated sand layer is much larger than that of the dry sand (see **Figure 7(a)**).

2.2. Dual particle jetting model

The decomposition of the particle shell into the inner compact and outer dilute layers as a result of shock interaction prompts us to speculate that the fragmentation of the inner and outer layers correspond to the primary and secondary particle jetting, respectively. This speculation satisfies some fundamental facts that (1) the primary and secondary particle jets initiate from the inner and outer surface of particle shells, respectively; (2) the secondary particle jetting occurs upon the reflection of the shock wave on the outer surface; (3) the primary jets overtake the primary jets in later times. Therefore, a dual particle jetting model illustrated in **Figure 8** has been put forward to account for the formation of the primary and secondary jets. The following task is to elaborate the proper models describing the respective fragmentation of the inner and outer particle layers. These models should be based on the underlying mechanisms and validated against the experimental results, the onsets of primary/secondary

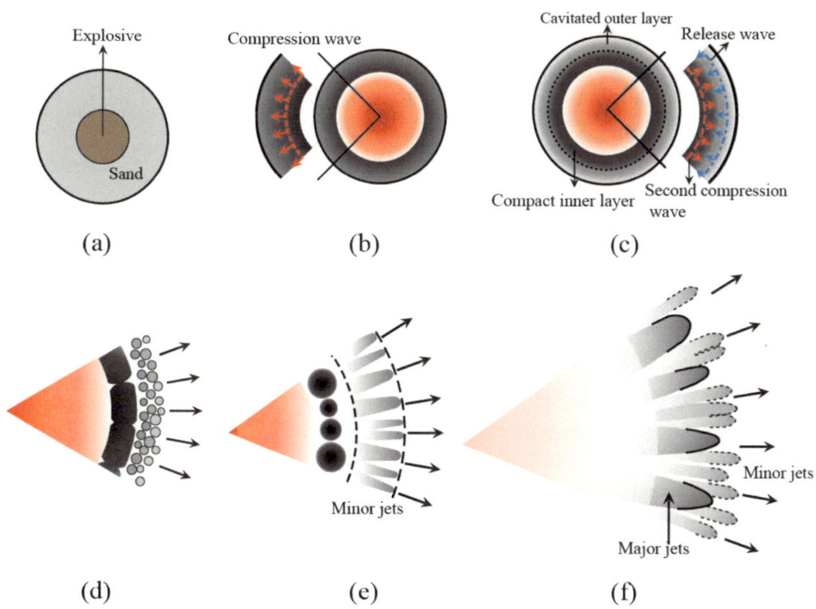

Figure 8. Illustration of the dual particle jetting model, which consists of the formation of the inner compact and outer dilute layers, and the breakup of these two distinct layers [22].(a):the initial annular configuration; (b): expansion of detonation gases issues the compression wave; (c): the reflected rarefaction wave causes the spallation of outermost layer; (d):fragmentation of the inner compact layer; (e): protrusion of secondary jets; (f): overtake of secondary jets by primary jets.

jetting, the size of primary/secondary jets, and the dependence of the jet number on a variety of factors as well.

2.3. Primary particle jetting model: the destabilization of the expanding shell

The consistent density and velocity across the thickness of the inner compact layer indicate that the compacted layer expands as the incompressible shell. Under this premise, we consider a sphere shell characterized by an inner radius R_1 and outer radius R_2 as shown in **Figure 9**, which can be determined by the hydrodynamic simulations (see **Figure 6**). The thickness of the shell is R_2–R_1. Adopting the spherical coordinate system associated with the frame (e_r, e_θ, e_ϕ), the outward divergent motion of the continuous sand shell demonstrated in experiments is modeled by applying a uniform velocity $Vr\, e_r$ at the inner surface ($R = R_1$), which can also be derived from the hydrodynamic simulation (see **Figure 7**).

Applying the continuity and momentum equations to the incompressible granular shells that can be described as viscoplastic materials, the analytical circumferential stress can be derived as follows (details of formulation can be referred to Ref. [13]).

$$\begin{aligned}\sigma_{\theta\theta} &= \frac{\rho V_1^2 R_1^4}{2}\left(\frac{1}{R^4}-\frac{1}{R_2^4}\right)+\frac{\eta V_1 R_1^2}{\sqrt{2}}\left(\frac{1}{R^3}+\frac{2}{R_2^3}\right)\\ &-2\rho(V_1^2 R_1 + A R_1^2)\left(\frac{1}{R}-\frac{1}{R_2}\right)+\sqrt{3}\,\tau_c\left(2\ln\frac{R}{R_2}+1\right),\end{aligned} \tag{1}$$

where ρ is the mass density of the sand shell, τ_c is the yield stress, and η is dynamic viscosity. Bearing in mind that the yield stress, τ_c, is a function of both saturation degree and the pressure applied on the inner surface which is in the order of O(10^0–10^1) Mpa (see **Figure 5(c)**), the yield stress of saturated sand (~1 MPa) is much lower than that that of dry sand (~13.7 MPa) due to the lubrication effect assuming average pressure $\bar{P}$~10 MPa. The dynamic viscosity, η, is in the order of O(10^{-1}).

To predict the instability onset of the expanding sand shell, we will invoke a criterion for instability that has been shown to reasonably emulate more rigorous stability analysis [28]. This method can be viewed as an application of Le Chatelier's principle that states that for a system to be stable any deviation from equilibrium must bring about forces that tend to restore equilibrium. In general, the loss of stability is assumed to take place when an increment in strain occurs with no simultaneous increase in pressure or in load.

To obtain the circumferential pressure in the expanding shell, the circumferential stress from Eq. (1) is integrated through the thickness h of the shell,

$$\begin{aligned}T &= \int_{R_1}^{R_2}\sigma_{\theta\theta}\,dR\\ &= \frac{\rho V_r^2 R_1}{2}\left(\frac{1}{3}-\frac{4R_1^3}{3R_2^3}+\frac{R_1^4}{R_2^4}\right)+\eta V_r\sqrt{2}\left[\frac{1}{4}+\frac{3R_1^2}{4R_2^2}-\frac{R_1^3}{R_2^3}\right]\\ &-2\rho(V_r^2 R_1)\left(\ln\frac{R_2}{R_1}-1+\frac{R_1}{R_2}\right)+\sqrt{3}\,\tau_c R_1\left(1-\frac{R_2}{R_1}-2\ln\frac{R_1}{R_2}\right).\end{aligned} \tag{2}$$

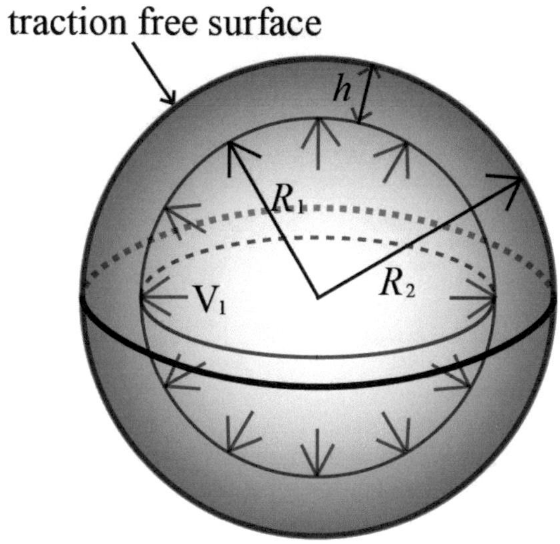

Figure 9. Configuration of an expanding spherical shell with inner radius R_1 and outer radius R_2 [13].

Figure 10(a) and **(b)** plot the variations of circumferential tension in dry and saturated sand shells with the expansion of the shell driven by the detonation of central TNT or HMX, respectively. The parameters are chosen as follows: $\rho = 2.1 \times 10^3$ kg/m^3, $V_{r,dry,tnt} = 220$ m/s, $V_{r,dry,hmx} = 330$ m/s, $V_{r,saturated,tnt} = 380$ m/s, $V_{r,saturated,hmx} = 420$ m/s, $\tau_{c,dry} = 15$ MPa, $\tau_{c,saturated} = 0.5$ MPa. The terms with the coefficient involving η can reasonably be ignored as a result of the dimensional analysis. The instability onset is identified as the point at which $dT/dR_2 = 0$, beyond which the increase of strain does not render the corresponding increase of the pressure or loads. Specifically, the critical radius of dry and saturated sand shells corresponding to the destabilization onset driven by detonation of TNT or HMX are $R_{c,dry,tnt} = 75$ mm, $R_{c,dry,hmx} = 80$ mm, $R_{c,saturted,tnt} = 98$ mm, $R_{c,saturted,hmx} = 105$ mm, respectively. Clearly, faster detonation velocity of explosive and addition of interstitial fluids can effectively delay the destabilization onset of the inner compact layer, equivalently the initiation of the primary jetting, consistent with the experimental observations. Likewise, we can predict the destabilization onsets of expanding sand shells with varying moisture contents as plotted in **Figure 11**, which agree well with those derived from the experimental observations. Note that the observed destabilization onsets of particle shells were determined from the high-speed photos that show the visible patterns in the surface of charge, which actually occurs after the destabilization onset.

The fragment size following breakup is substantially determined by the wavelength of the most unstable disturbance that has the greatest growth rate. Determination of a dominant unstable wave length is difficult due to the time-varying nature of the mean flow. Louis suggested that for a small value of Γ, the most disturbances are in a range of wavelengths between $O(1)$ and $O(1/\Gamma)$ times the instant thickness of the shell, where Γ is the dimensionless number as follows

$$\Gamma^2(t) = \frac{6\rho\, \bar{d}^{\,2}_{\theta\theta}(t)\, h^2(t)}{\tau_c}. \tag{3}$$

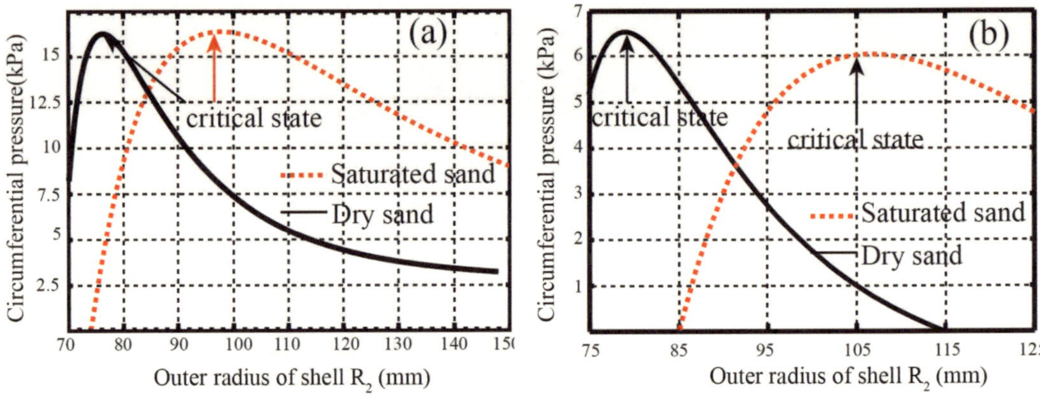

Figure 10. The variations of circumferential tension in dry and saturated sand shells with the expansion of the shell driven by the detonation of central TNT (a) or HMX (b).

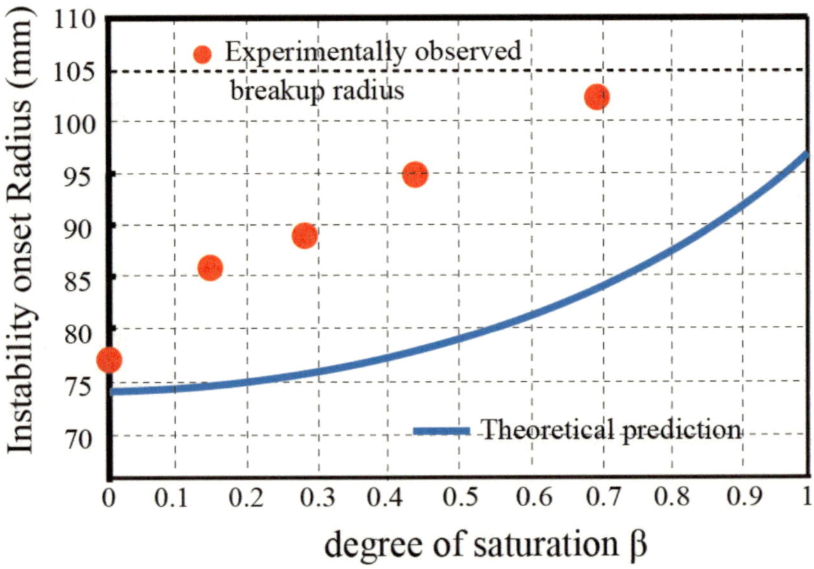

Figure 11. Theoretically predicted (curve line) and experimentally observed (red circles) critical radii of expanding sand shells with varying saturation degree driven by the detonation of central TNT.

In Eq. (3), $\bar{\varepsilon}_{\theta\theta}$ and h are average circumferential strain and the instant shell thickness, respectively. Along this line, we give the first order of the estimation of the range of fragment mass of dry sand shells, or equally jet mass m_{jet}, as a function of the yield stress as shown in **Figure 12**. The experimental determined jet mass for the dry and moderately wetted sand falls well into the range predicted by the aforementioned model.

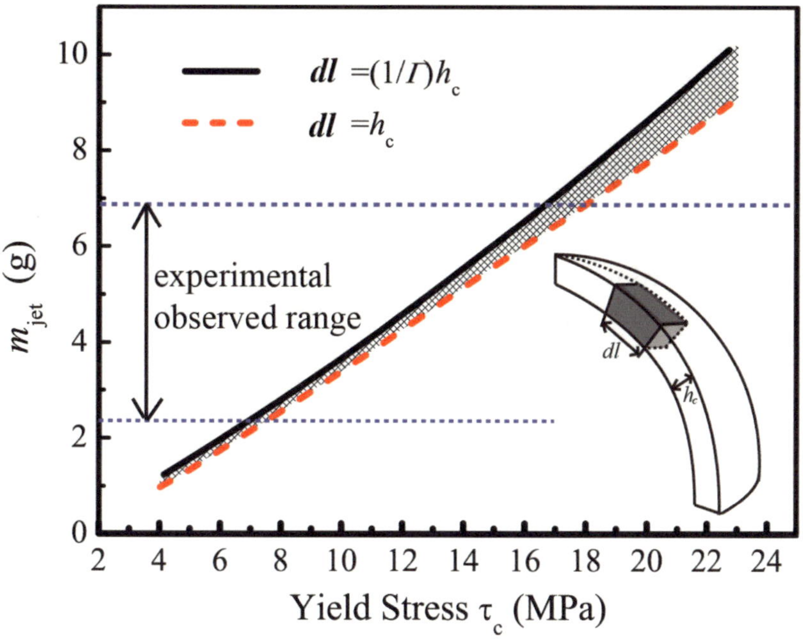

Figure 12. Primary jet mass vs. yield stress of sand shells [13].

2.4. Secondary particle jetting model: cavitation model based on the expansion of hollow spheres

A micromechanical approach describing the cavitation process originally applied to ductile damage in solids has been proposed to account for the spallation in a liquid (or melt metal) subjected to a pulsed tensile load [29]. Xue et al. [22] adapted this cavitation-based spallation model to account for the disintegration of the outer particle layer, or equivalently, the formation of the secondary particle jetting, which is initiated by the unloading wave opposed to the tensile loading.

The incipient spallation of the outer particle layer takes the form of the macroscopic dilation in the wake of rarefaction waves. The dependence of the volumetric variation on the pressure is schematically plotted in **Figure 13(a)**. Within the frame of cavitation model, the bulk of the sample is seen as a collection of adjacent hollow spheres of internal and external radii $a(t)$ and $b(t)$ (see **Figure 13(b)**), respectively. The initial outer radius of the sphere b_0 can be interpreted as the mean half-length between two neighboring nucleation sites as depicted in **Figure 13(b)**. As b_0 defines the mass volume involved in the cavitation pattern, the "microscopic" pressure invoked by the cavitation varies with b_0. Since the "microscopic" pressure should agree with the "macroscopic" pressure dictated by the volumetric variation, this compatibility provides a criterion for the determination of b_0. To ensure the expansion of the microscopic hollow sphere is compatible with the dilation of the macroscopic outer particle layer, the microscopic expansion rate of the sphere, $3\dot{b}/b$, should remain consistent with the macroscopic dilation rate of the particle layer, $\dot{V}/V$, where V is the volume of the outer particle layer. The dilation rates at these two length scales are thereafter detonated by a single parameter D.

The spallation or, equally, the dilatation process of the outer layer consists of three stages. The first so-called hollow sphere expansion stage is prescribed by the relaxation of the accumulated pressure when the volumetric increase is dictated by the dilatation rate D. During the phase I, voids hardly begin to grow due to the inertial resistance. The end of the phase I of cavitation coincides with the full relaxation of the pressure marked by the restoration of the initial packing density. Afterward, the rapidly expanding matrix progressively becomes gaseous so that the particles interact by collision and the continuous displacement/stress field does not exist. Thus, the matrix and the void of the hollow sphere undergo the independent inertial expansion. The gaseous regime of the matrix is hereinafter detonated as the phase II of cavitation, which sustains as long as the matrix remains diluter than the initial packing state. Examining the packing density of the matrix in the dry sand suggests that the gaseous state of matrix maintains even when the fragmentation starts. By contrast, the gaseous saturated sand is soon transformed to the dense granular flows when the loose particles get recompressed

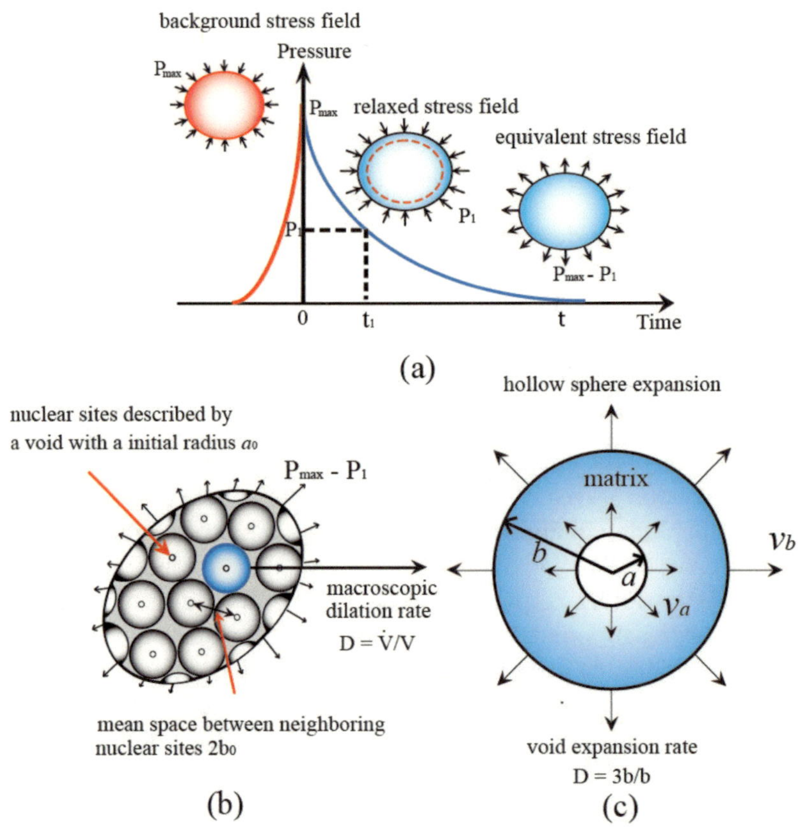

Figure 13. (a) Pressure relaxation experienced by the dry (dashed line) and saturated (solid line) particle layer accompanied by the dilation; (b) schematic of the hollow sphere pattern; (c) the expansion of the individual hollow sphere [22].

by the unconstrained outward expansion of the void. The subsequent expansion of the void, detonated as the phase III, is conditioned by the dense granular flow in the incompressible matrix.

Analytical modeling of these three sequent phases can be referred to Ref. [22]. This cavitation model estimates that the fragment size or equally the secondary jet size for dry and saturated sand ranges from 4 to 6 mm and 1.6 to 3.3 mm, respectively. Applying the proper fragmentation criterion, the predicted onset of secondary particle jetting occurs at 200–300 μs for the dry sand and 50–100 μs for the saturated sand after the detonation, respectively. The cavitation model is capable of predicting the fragmentation onset and the fragment size consistent with the experimental results. Therefore, cavitation is inferred here to be the most probable spallation mechanism of the outer particle layer.

The size of the secondary jets represented by twice the length between two activated nucleation sites, $2b_0$, is dictated by the compatibility of the "microscopic" and macroscopic pressures during the unloading of the compacted particles. Mathematically, smaller b_0 in saturated sand is rendered by the significantly elevated dilation rate due to the larger elastic energy and faster moving release waves in the saturated sand. Micromechanically, it is the results of the competition between two neighboring cavities. Analogous to the scenario involving the Mott waves traveling between fractures (see **Figure 14(a)**), the expansion of cavity emanates the compressive waves into the neighborhood so as to suppress the potential cavitation nucleation in the encompassed area. The combined travel length of the compression waves emanating from the neighboring nucleation sites can be taken as the upper limit of the spacing between nucleation sites, namely $2b_0$ (see **Figure 14(b)**). The unloading duration in saturated sand is almost one order shorter than that in dry sand, leading to the significantly shortened distance between two neighboring cavities.

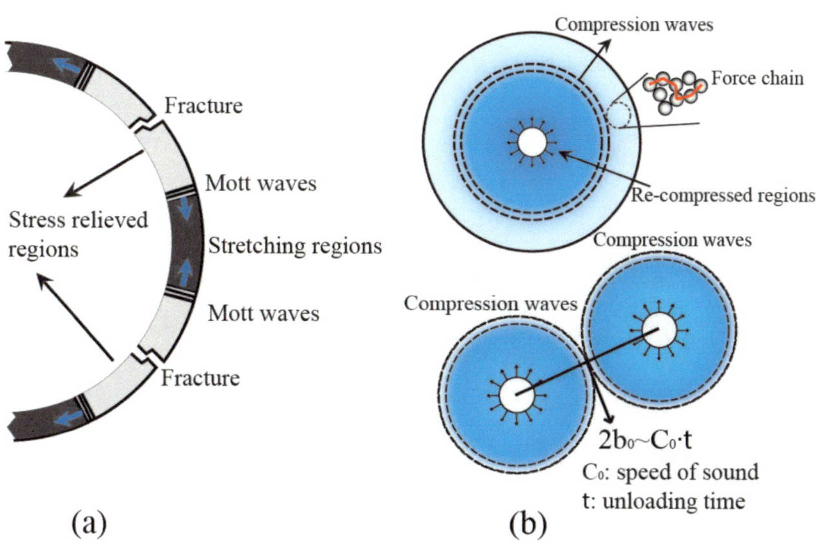

Figure 14. (a) A schematic of the Mott cylinder model with regard to the dynamic fragmentation of the solid cylinder (ring); (b) in particles, compression waves propagate away from an activated nucleation sites (above) retarding any activation of the nucleation sites within the travel radius and collide with those emanating from the adjacent nucleation sites [22].

3. Shock-induced particle jetting

3.1. Quasi-two-dimensional particle jetting under moderate impulsive loads: phenomenal description

It is difficult to visualize the particle jet spread in the spherical or cylindrical experiments of the explosive-driven particle jetting due to the superimposition of jets, obscured by detonation gases, and the very short timescale as well. To overcome these disadvantages, Rodriguez et al. [1, 2] studied the particle jetting in quasi-two-dimensional configurations using moderate pressure loads induced by shock-tube-type facilities connected to a Hele-Shaw cell. With this convenient experimental setup, it is possible to conduct repetitive reliable experiments using a ring of particles in radial expansion trapped in a Hele-Shaw cell as shown in **Figure 15(a)**. More importantly, it is much easier to visualize and distinguish the primary and secondary jets. Xue et al. carried out the experiments of the shock-induced particle jetting using the apparatus similar to that devised by Rodriguez and reported similar observations of the particle jetting process.

Figure 15(b) shows the evolutions of dual particle jets of flour ring dispersed by the shock wave with the overpressure of 3.33 bar. The perturbation of the inner surface of ring can be detected at t = 1 ms. The primary jets cutting through the inner surface are well defined in

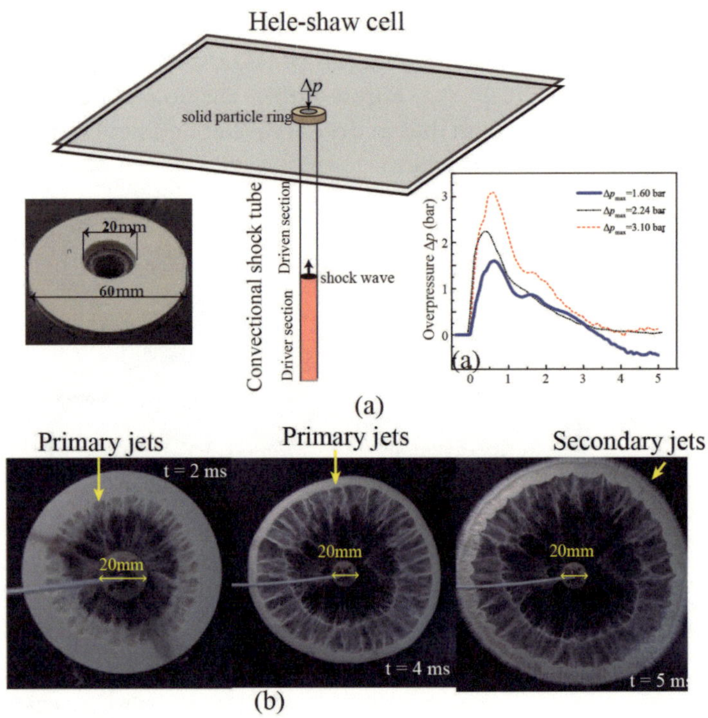

Figure 15. (a) Schematic of the quasi-two-dimensional experimental setup for shock induced particle jetting. Insets: photo of the four ring sample (left) and the overpressure histories at the exit of shock tube (right). (b) High speed photos of particle induced particle jetting.

the first several milliseconds. A large number of secondary jets burst out of the outer surface of ring 1.5 ms after the shock front reaches the outer surface. Afterward, the needle-like secondary jets undergo dramatic growth during the following one millisecond, while the tips of primary jets seem to be arrested at the bottom of secondary jets. It takes another several milliseconds that the primary jets overtake the secondary jets.

3.2. Particle-scale evolution of shock-induced particle jetting: DEM investigation

Experimental observations can only provide the configurational evolution of particle ring having no access to the particle-scale information, such as the particle velocities and forces. DEM has proven to be an effective tool to investigate the particle-scale velocity and stress fields in particles subjected to the static or dynamic loadings. Xue et al. performed the DEM simulations of the shock-induced particle jetting using the same geometrical configuration as in the experimental. Parametric studies were carried out to quantify the effect of a variety of variables, including the overpressure of shock loading (p_0), the inner and outer radii of ring (R_{in} and R_{out}), the packing density (χ), and particle size (d_p). Details of the simulation can be found in Ref. [25].

Figure 16 shows the shock dispersal of particle rings in terms of variations in velocity profiles. The shock-loaded particle rings with different initial parameters develop into the resembling jet structures with distinctive features as demonstrated in **Figure 16**. The formation and evolution of the primary jets in all cases, which are barely accurately described using experimental techniques, undergo two distinctive phases, namely the nucleation of the incipient jets and the competitive growth of the incipient jets. Here, the incipient jets are referred to as the localized shear flows or, equivalently, the fast moving particle clusters as shown in the innermost frame in each subfigure of **Figure 16**. The inner surface of ring remains smooth without visible dents or ripples so that the first phase is almost impossible to identify from the experimental observations.

The azimuthal velocity profiles of particle ring in early times shown in **Figure 17** demonstrate the nucleation of the incipient jets. No consistent pattern persists during the first millisecond, the spikes in the azimuthal velocity profile being transient and irregular. The flows behind the shock front are largely homogeneous around the perimeter. The following several milliseconds saw the dramatic transition of azimuthal velocity profile from irregular oscillations to regular fluctuations that are consistent throughout. This transition is clearly manifested by the substantial jump around t = 0.5–1 ms in the variations of correlation coefficient of the two sequential azimuthal velocity profile (see **Figure 17(b)**). The peaks indicated in **Figure 17(a)** correspond to the localized shear flows, or equivalently the incipient jets identified in **Figure 17(d)**.

The radial growth of incipient jets in terms of the penetration depth into the bulk and the cross-sectional width is strongly uneven, the strong jets mushrooming outwards opposed to the retarded weak jets. As a result, the substantial elimination and the coalescence of weak jets prevail throughout the second phase. By contrast, the mushroom-like strong jet occasionally would split into multiple subjets, which is more likely to occur in rings with low packing density (see **Figure 16(c)** and **(d)**). Interestingly, the multiplication of strong jets can take place multiple times. The evolutional characteristics of incipient jets revealed by the DEM simulations are substantiated by the experimental observations (see **Figure 18**).

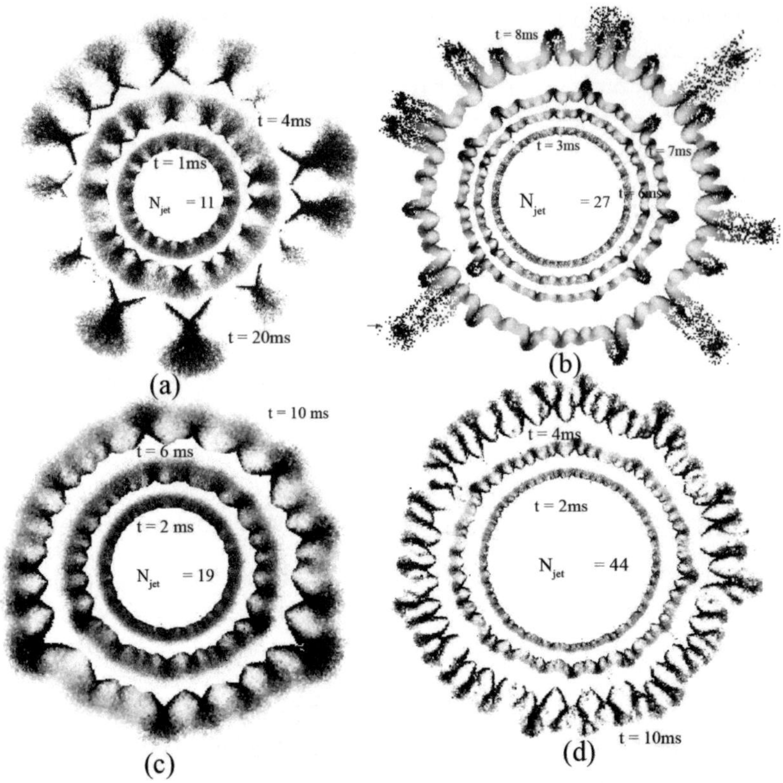

Figure 16. Evolutions of the velocity profiles in particle rings with different parameters. Particles are shaded according to the magnitude of velocities. (a) $d_p = 2$ mm, $p_0 = 5$ bar, $R_{in} = 20$ cm, $\chi = 0.55$; (b) $d_p = 2$ mm, $p_0 = 5$ bar, $R_{in} = 35$ cm, $\chi = 0.55$; (c) $d_p = 2$ mm, $p_0 = 5$ bar, $R_{in} = 20$ cm, $\chi = 0.42$; (d) $d_p = 1$ mm, $p_0 = 5$ bar, $R_{in} = 20$ cm, $\chi = 0.45$.

The elimination of the weak jets significantly influences the temporal variations of the jet number as shown in **Figure 19**. After the chaotic initiation of incipient jets during the first several milliseconds evidenced by the strong oscillation of jet number, the jet number plummets dramatically in the following 5–10 ms. Afterward, the jet number undergoes much more gradual decrease until the jets are expelled from the outer surface of ring. Taking into account these fundamentals demonstrated in **Figure 19**, a physics-based equation as follows can be derived to describe the temporal variation of jet number, N_{jet}.

$$N_{\text{jet}} = N_{\text{jet},i}\left(R_{\text{in}}, d_p, \chi\right) - V_{\text{jet}}(p_0, \chi)\Delta t(h, p_0, \chi). \tag{4}$$

In Eq. (4), $N_{jet,i}$ represents the number of initial activated incipient jets; V_{jet} represents the decline rate of jet number during phase II (number per unit time); Δt is the duration of phase II. Surprisingly, the overpressure of shock waves does not have the discernible effect on the number of initial jet, $N_{jet,i}$, which instead is a function of the inner radius of ring, R_{in}, the particle diameter, d_p, and the packing density, χ. It suggests that $N_{jet,i}$ is indicative of some intrinsic characteristics of particles, analogous to the intrinsic flaws of solids. The decline rate of jet number, V_{jet}, is clearly elevated by stronger shock loadings. Besides, lower packing density seems to hinder the elimination of jets. The duration of phase II, Δt, is among the most important factors governing the jet

http://dx.doi.org/10.5772/intechopen.68795

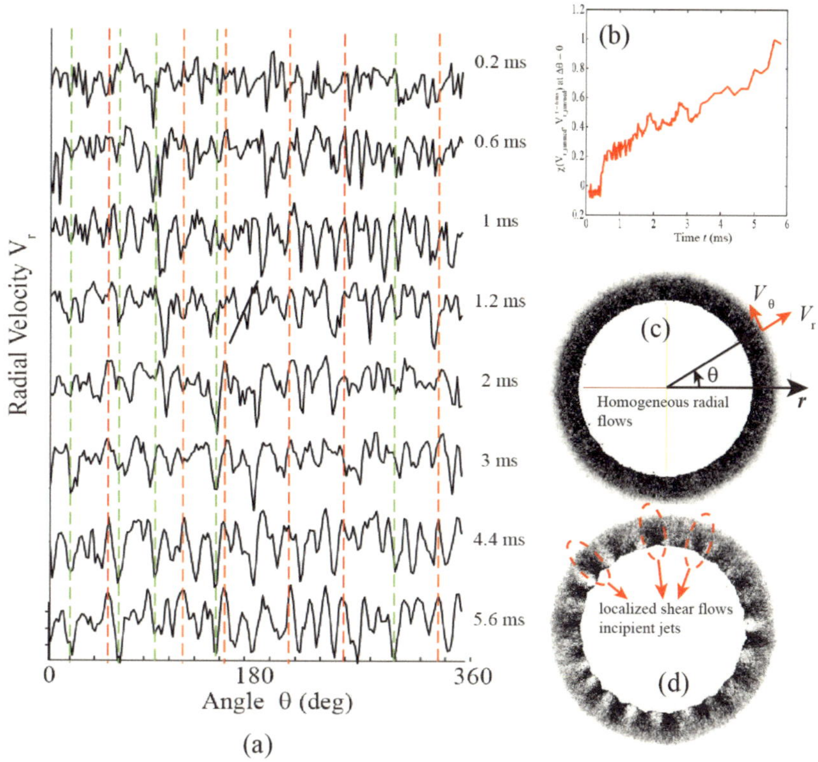

Figure 17. (a) Azimuthal velocity profiles of particle ring in early times and (b) variations of correlation coefficient of the two sequential azimuthal velocity profile of particle ring. Snapshots of particle ring at t = 0.2 ms (c) and t = 3 ms (d).

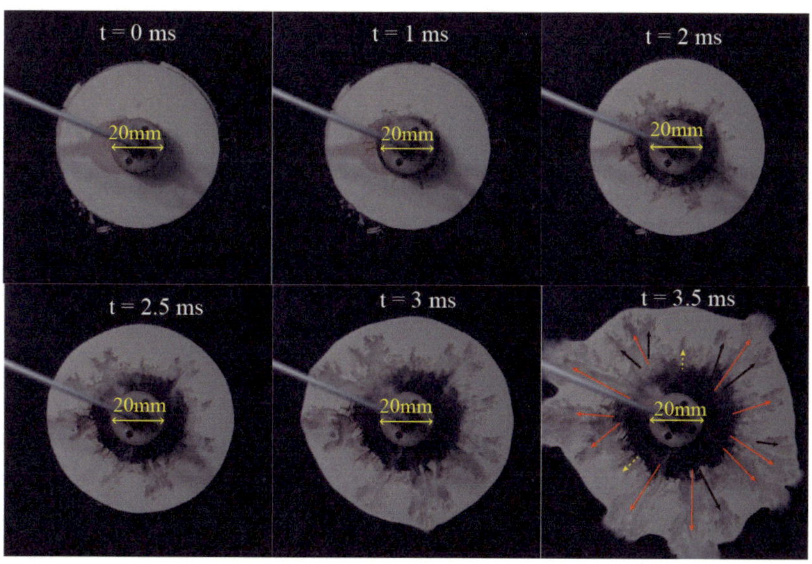

Figure 18. High speed photos of shock dispersal of corn quartz sand ring.

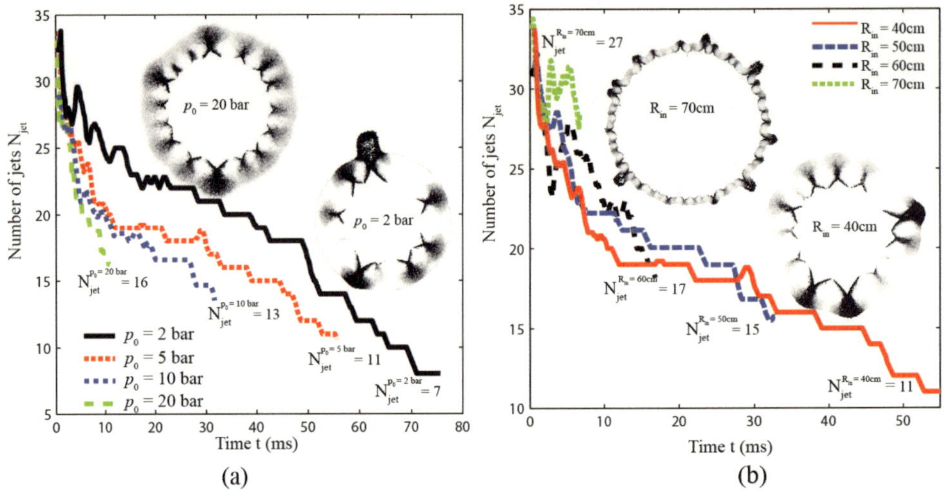

Figure 19. Temporal variations of the jet number with varying overpressure of shock loadings (a) and inner radius of ring (b). The inner radius of ring remains constant, R_{in} = 20 cm in (a). The overpressure peak in (b) is 5 bar.

number, since the significant increase of jet number either due to the stronger shock loadings or larger inner radius of ring is dominantly caused by the truncated phase II. In another way, there is not enough time for the elimination of jets to fully unfold. The thickness of ring, h, the overpressure of shock loadings, p_0, and the packing density, χ, are among the parameters influencing Δt.

3.3. Mechanisms governing the shock-induced particle jetting

The analytical formulation of Eq. (4) entails a thorough understanding of the underlying mechanisms, specifically the formation and elimination mechanisms of incipient jets. With regard to the formation of incipient jets, it is necessary to unlock the transition of the homogeneous flows to the localized shear flows. Unlike solids or liquids, the stress waves in particles travel through particle contact points and are primarily transmitted by the "force chains" that carry most of load in the granular materials [18, 30]. Meanwhile, the shock energy is dissipated by the random particle collisions. Because of the strong energy dissipation and nonlinear characteristics of granular systems, the inter-particle forces are transmitted through heterogeneous architecture of force chains such as shown in **Figure 20**, where the inter-particle contact forces are represented by inter-particle lines scaled with the magnitude of the contact forces. The initial contact network of particles (see the top panel in **Figure 20**) appears to be homogeneous in general with particle-scale heterogeneities. The cylindrical shock loading activates the contacts aligning with the local radial directions. Besides the intricate contact network in the innermost particle layers, a handful of long linear force chains extend radially from the inner surface toward the outer surface (shaded red in the second panel in **Figure 20**). These long linear force chains act as the arteries from which a growing number of short force chains are initiated, forming distinguishable clusters of force chains at t = 1 ms with the dimensions much larger than that of constituent particles.

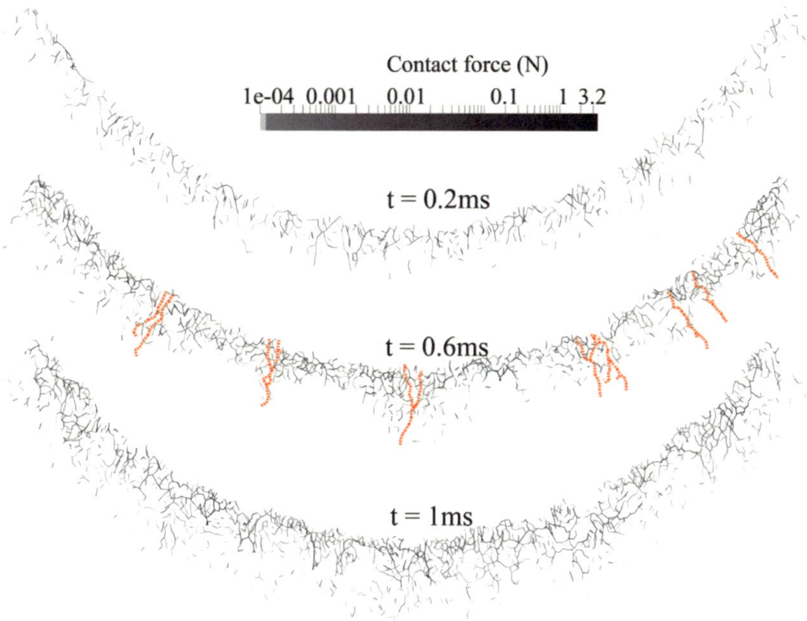

Figure 20. Snapshots of the network of force chains in the bottom section of the particle ring subjected to the shock loading of P_0 = 20 bar in early times. Force chains (denoted by the thick dashed red lines)at t = 0. 6 ms indicate the long linear force chains acting as the nuclei of the force chain clustering [25].

The variations in the circumferential distributions of strong contact density, $\rho_{contact}$, in early times (see **Figure 21**), demonstrate how the particle-scaled heterogeneities evolve into the macroscale clusters of strong contacts indicated by the contact force peaks with width much larger than the particle size. Note that the agglomeration of force chains is well ahead of the formation of the nonuniform velocity profile that signifies the beginning of the particle clustering. Since the momentum alongside the stresses is being transmitted along the force chains, leaving the particles disconnected from the force chains, there are few chances to obtain the momentum. Particles connected by the strong force chains are supposed to move faster than those cut off from the contact network. Force chains thus act as the main channels of momentum at least in early times as suggested by the strong correlation between the Azimuthal distribution of contact density $\rho_{contact}$ and radial velocity V_r in the first millisecond as shown in **Figure 21(a)**.

Force chains also play a major role in the elimination of weak jets caused by the dilating strong jets as demonstrated in **Figure 22**. With the incipient jets (composed of the red circles in **Figure 22**) moving ahead of the slow-moving particles (denoted by the green-dashed circles in **Figure 22**), velocity differences across the edges of the incipient jets retard any sustained contacts, leading to the weakened lateral confinement imposed on the jets. Therefore, nontrivial transverse flows occur along the edges of jet, the jet front flaring out significantly (see the middle panel in **Figure 22**). The lateral expansion of adjacent jets, especially the jet heads, squeezes the slow-moving particles in between (denoted by blue-dotted circles in **Figure 22**) establishing an intricate network of force chains therein

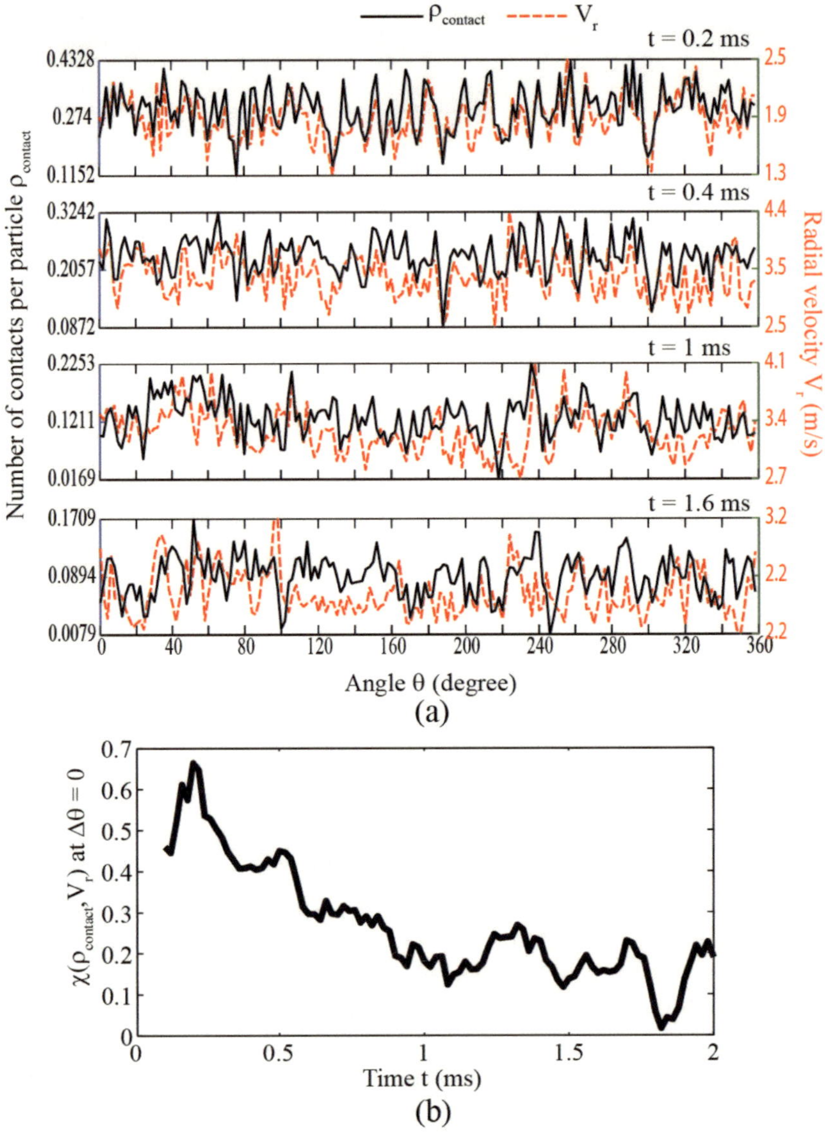

Figure 21. (a) Azimuthal distribution of contact density $\rho_{contact}$ and radial velocity V_r of particle ring at different times and (b) temporal evolution of correlation coefficient between $\rho_{contact}$ and V_r.

(see the middle and bottom panels in **Figure 22**). The newly constructed force chains with the dominant transverse orientation hinder the radial transport of the momentum that is instead channeled along the transversely aligned force chains (see the middle panel in **Figure 22**). The growth of the burgeoning minor jets between two major jets is thus likely to be suppressed or even retarded. The minor jets composed of particles indicated by the dotted circle in the middle panel of **Figure 22** are degraded to the slow-moving cluster. With the slow-moving particles increasingly lagging behind, more spaces are left outside the edges of jets, resulting in the intensified transverse flows along the edges. By contrast, the radial compaction leads to the enhanced radial resistance restraining the radial advance of the jet front such as illustrated in the bottom panel of **Figure 22**. At some point, the transverse flows along the edges of jets are expected to overwhelm the radial propagation. The edges of major jets curl outward toward the opposite directions so that the major jet splits into several subjets (indicated by the circles in the bottom panel in **Figure 22**). The subjets with the propagation direction deviating from that of the parental jet would undergo the same development described above until they are expelled from the outer surface.

Given that the suppression of weak jets by the strong jets is mainly responsible for the decrease of jet number, the decline rate of jet number, V_{jet}, decidedly depends on the spatial density of

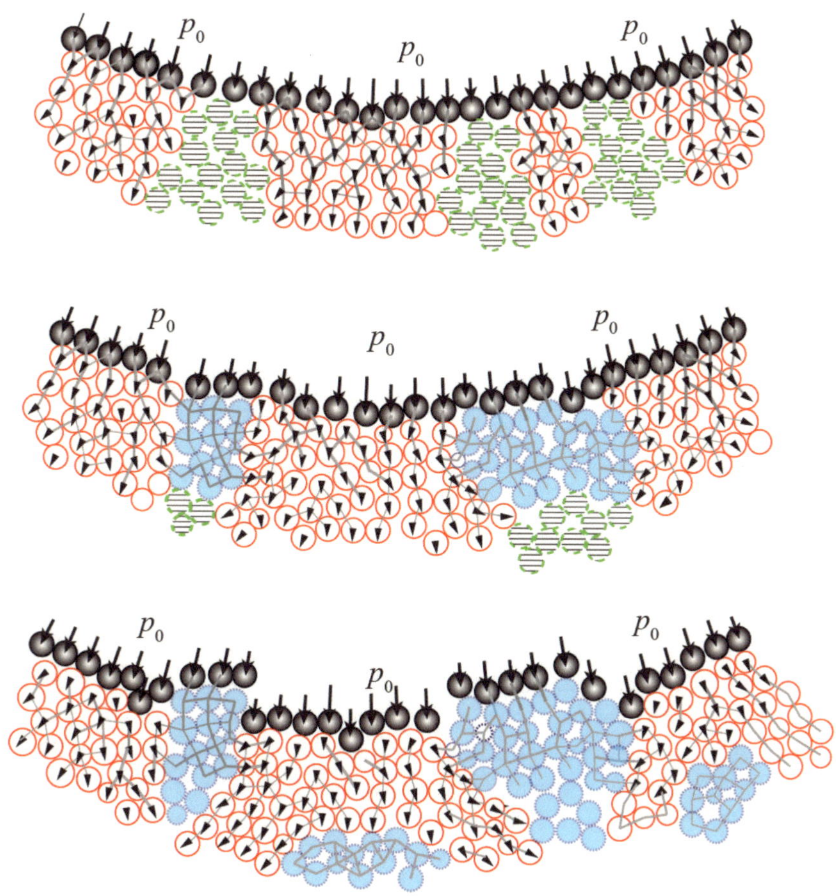

Figure 22. Illustrations of the evolution of the jetting pattern as well as the contact network. The red circles, dashed-line filled circles and blue filled circles represent the fast-moving particles connected by force chains, slow-moving particles without effective contacts among them, and slow-moving particles connected by transversely oriented force chains, respectively [25].

incipient jets, the perimeter of ring, and the transverse expansion of strong jets. The average spacing of initial jets varies little with R_{in} and p_0, whereas decreases with decreasing packing density and particle size. The transverse expansion of strong jets strongly correlates with the radial propagation of jets that are driven by the impulsive loadings. Accordingly, stronger shock loading intensifies both the radial and transverse expansion of jets, hastening the suppression of the adjacent weak jets.

Figure 23 highlights the key events characterizing the formation and competitive growth of incipient jets. An excessive large number of strong force chains extruding into the bulk serves as the nuclei of incipient jets. The jets born earlier or showing stronger shear flows undergo considerable transverse flare up, annihilating the burgeoning weak jets. A substantial portion of initial incipient jets cannot survive the first instants of the phase II.

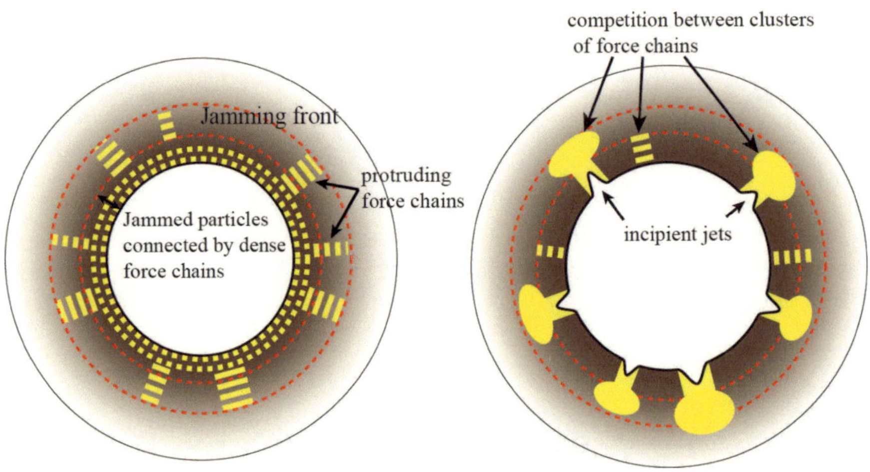

Figure 23. Illustration of key events dominating the formation (left) and elimination (right) of incipient jets.

4. Discussion

Despite the resembling jetting pattern driven by the central explosion and radial shock loadings, the underlying mechanisms are fundamentally different as required by the distinct behaviors of particles subjected to strong blast waves and modest shock waves. In the former case, particle layers are compacted so tightly that they expand like the solids of the constituent materials. Thus, the (primary) particle jetting may well be understood from the continuum perspective since the hydrodynamic instability of interface, such as RT instability, fails to predict the jetting timescale comparable with the experimental data. Bulk fracture of compacted expanding particle layers becomes the promising candidate. The dynamic fragmentation theories of solids may well be applicable to the theoretical model of the explosion-driven particle jetting. But some major alterations need to be made to adapt these theories to the fragmentation of particle assemble. Since particles cannot sustain the tensile stresses nor have the surface energy, the fracture criterion of solids involving these two pivotal variables does not hold in particles. Experimental results suggest that the inception of particle jetting initiates shortly after the propagation of the rarefaction wave. This observation implies that the unloading of particles triggers the particle jetting. The fractures of solids mainly nucleate at the intrinsic flaws that determine the statistics of fragment size. By contrast, the dimension of flaws in particle system, namely the inter-grain pores, contradicts with that of particle jets. It is plausible to assume that the nuclei of the particle jets may well be brought in by a strong shock interaction. The implosion of particles causes dozens of shear bands across the thickness of particle ring with attrited grains [31]. Recent experiments also collected the sintered clumps of aluminum powders after the explosive dispersal of powders [32]. The heterogeneous thermodynamic activities occurring in the blast loaded particles, such as shear banding, should be the focus of the future study. A thorough understanding in this regard needs the rigorous examination of previous experimental data and development of adequate experimental and numerical techniques providing more direct evidences.

Shock-induced particle jetting opens a fundamentally different domain but attracts relatively less attention compared with the explosion-driven particle jetting. This scenario offers an ideal opportunity to look into the transient particle flows. This chapter presents some preliminary investigations into this problem via both experimental and numerical methods. Much more work is needed to clarify the origin of the nuclei of incipient jets and the interplays between jets with varying strengths. In this regard, force chains play an essential role via introducing the inhomogeneity and modulating the jetting pattern.

5. Conclusion

Both explosion-driven and shock-induced particle jetting exhibit the dual jetting structure, namely, the primary jets initiating from the inner surface and the secondary jets initiating from the outer surface of particle rings/cylinders/shells. The primary and secondary jets have fundamentally different size and occur in different times so that respective mechanisms are required. More importantly, distinct behaviors of particles subjected to strong blast waves and weak shock waves dictate different mechanisms underpinning the particle jetting in both cases. Accordingly, we adopt a continuum approach to model the explosion-driven particle jetting. Specifically, a destabilization model of expanding shell is proposed to account for the onset of the primary jetting. The secondary jetting can be described by a cavitation spallation model based on the expansion of hollow spheres. The timescale and characteristic size of primary/secondary jets predicted by theoretical models agree well with the experimental data. By contrast, the shock-induced particle jetting is studied via the DEM method, which can access the particle-scale information, such as particle velocities and contact forces. The investigation reveals a two-staged evolution of particle (primary) jets, the formation and competitive growth of incipient jets. The formation of incipient jets is characterized by the transition from the homogeneous flows to the localized shear flows. The ensuing evolution of incipient jets is accompanied by the substantial annihilation of weak jets and the multiplication of strong jets. The mechanisms underlying these two phases are found to be closed related with the network of force chains.

Author details

Kun Xue*, Xiaoliang Shi, Kaiyuan Du and Haoran Cui

*Address all correspondence to: xuekun@bit.edu.cn

Key State Laboratory of Explosive Science and Technology, Beijing Institute of Technology, Beijing, China

References

[1] Rodriguez V, Saurel R, Jourdan G, Houas L. External front instabilities induced by a shocked particle ring. Physical Review E. 2014;**90**:043013

[2] Rodriguez V, Saurel R, Jourdan G, Houas L. Solid-particle jet formation under shock-wave acceleration. Physical Review E. 2013;**88**(6):063011

[3] Ripley RC, Zhang F. Jetting instability mechanisms of particles from explosive dispersal. Journal of Physics: Conference Series. 2014;**500**(15):152012

[4] Ripley R, Donahue L, Zhang F. Jetting instabilities of particles from explosive dispersal. AIP Conference Proceedings. 2011;**1426**(1):1615-1618

[5] Milne AM, Floyd E, Longbottom AW, Taylor P. Dynamic fragmentation of powders in spherical geometry. Shock Waves. 2014;**24**(5):501-513

[6] Milne A, Parrish C, Worland I. Dynamic fragmentation of blast mitigants. Shock Waves. 2010;**20**(1):41-51

[7] Grégoire Y, Sturtzer M-O, Khasainov BA, Veyssière B. Cinematographic investigations of the explosively driven dispersion and ignition of solid particles. Shock Waves. 2014;**24**(4):393-402

[8] Gregoire Y, David F, Oren P. Development of instabilities in explosively dispersed particles. AIP Conference Proceedings. 2011;**1426**(1):1623-1626

[9] Fue-Sang L, Tao X, Fan Z. The role of vorticity and turbulence on the instability of a dense solid particle flow. AIP Conference Proceedings. 2011;**1426**(1):1619-1622

[10] Frost DL, Loiseau J, Marr BJ, Goroshin S. Particle segregation during explosive dispersal of binary particle mixtures. AIP Conference Proceedings. 2017;**1793**(1):120020

[11] Frost, D.L., Gregoire, Y., Goroshin, S., and Zhang, F., "Interfacial instabilities in explosive gas-particle flows," Proceedings of 23rd International Colloquium on the Dynamics of Explosions and Reactive Systems, Univ. of California, Irvine, July 24-29, 2011

[12] David LF, Yann G, Oren P, Samuel G, Fan Z. Particle jet formation during explosive dispersal of solid particles. Physics of Fluids. 2012;**24**(9):091109

[13] Xue K, Li F, Bai C. Explosively driven fragmentation of granular materials. The European Physical Journal E. 2013;**36**(8):1-16

[14] Zhang F, Frost DL, Thibault PA, Murray SB. Explosive dispersal of solid particles. Shock Waves. 2001;**10**(6):431-443

[15] Zhang F. Metalized heterogeneous detonation and dense reactive particle flow. AIP Conference Proceedings. 2011;**1426**(1):27-34

[16] Ritzel D., Ripley R., Murray S., Anderson J.: Near-field blast phenomenology of thermobaric explosions. In: Hannemann K., Seiler F. (eds) Shock Waves(2009);305-310. Springer, Berlin, Heidelberg

[17] Frost DL, Ornthanalai C, Zarei Z, Tanguay V, Zhang F. Particle momentum effects from the detonation of heterogeneous explosives. Journal of Applied Physics. 2007;**101**(11): 113529

[18] Clark AH, Kondic L, Behringer RP. Particle scale dynamics in granular impact. Physical Review Letters. 2012;**109**:238302

[19] Zhang F, Ripley RC, Yoshinaka A, Findlay CR, Anderson J, von Rosen B. Large-scale spray detonation and related particle jetting instability phenomenon. Shock Waves. 2015;**25**(3):239-254

[20] Zarei Z, Frost D. Simplified modeling of blast waves from metalized heterogeneous explosives. Shock Waves. 2011;**21**(5):425-438

[21] David LF, Fan Z, Susan M, Stephen BM, Andrew JH, Marta S, Marc C-C, Chayawat O. Near-field impulse effects from detonation of heterogeneous explosives. AIP Conference Proceedings. 2002;**620**(1):946-949

[22] Xue K, Yu Q, Bai C. Dual fragmentation modes of the explosively dispersed granular materials. The European Physical Journal E. 2014;**37**:88

[23] Xu T, Lien F-S, Ji H, Zhang F. Formation of particle jetting in a cylindrical shock tube. Shock Waves. 2013;**23**:619-634

[24] Frost DL, Goroshin S, Ripley RC, Zhang F. Jet formation during explosive particle dispersal. City. 2010

[25] Xue K, Sun L, Bai C. Formation mechanism of shock-induced particle jetting. Physical Review E. 2016;**94**(2):022903

[26] Grujicic M, Pandurangan B, Qiao R, Cheeseman BA, Roy WN, Skaggs RR, Gupta R. Parameterization of the porous-material model for sand with different levels of water saturation. Soil Dynamics and Earthquake Engineering. 2008;**28**(1):20-35

[27] Grujicic M, Pandurangan B, Cheeseman BA, Roy WN, Skaggs RR. Application of the modified compaction material model to soil with various degrees of water saturation. Shock and Vibration. 2007;**14**:1-15

[28] Borg JP, Grady D, Cogar JR. Instability and fragmentation of expanding liquid systems. International Journal of Impact Engineering. 2001;**26**(1-10):65-76

[29] Huang X, Ling Z, Dai LH. Cavitation instabilities in bulk metallic glasses. International Journal of Solids and Structures. 2013;**50**(9):1364-1372

[30] Tai Q, Sadd MH. A discrete element study of the relationship of fabric to wave propagational behaviours in granular materials. International Journal for Numerical and Analytical Methods in Geomechanics. 1997;**21**:295-311

[31] Nesterenko VF, Meyers MA, Chen HC. Shear localization in high-strain-rate deformation of granular alumina. Acta Materialia. 1996;**44**(5):2017-2026

[32] Frost DL, Loiseau J, Goroshin S, Zhang F, Milne A, Longbottom A. Fracture of explosively compacted aluminum particles in a cylinder. AIP Conference Proceedings. 2017;**1793**(1): 120019

Section 4

Solid State of Granular Materials

Chapter 6

Particle Migration Phenomena Related to Hydromechanical Effects at Contact between Different Materials in Embankment Dams

Francesco Federico

Additional information is available at the end of the chapter

http://dx.doi.org/10.5772/67785

Abstract

The compatibility between a fine grained base (B) material and a downstream coarser particulate transition (T), under the seepage forces related to hydraulic gradients, plays a key role in safety of earthfill dams. This aspect, that is the possible migration of fine grains through the voids larger than their size, is analyzed according to a numerical procedure simulating the 1D-coupled particle migration and seepage unsteady states. The procedure accounts for grain size curve, constriction sizes and porosity of the materials as well as the rate of the suspension and drag forces associated with the seepage flow; friction triggered by normal contact forces induced by confining pressure is considered too. The procedure has been systematically applied to: (i) simulate the newly formed filter (F) at the contact of different B-T systems, (ii) review the criteria proposed by Terzaghi, and (iii) analyze the particle migration phenomena that affected some embankment dams.

Keywords: internal erosion, contact between base and transition materials, suffusion, numerical procedure

1. Introduction

Internal erosion phenomena are an important safety issue for embankment dams, dikes, and levees as shown by historical failures and incidents [1]. These phenomena [1] occur when soil particles are dragged by seepage in embankments (e.g., at the contact between core and transition or downstream materials), foundations, from embankment to foundation, around and into conduits through embankments and adjacent walls supporting embankments.

Four mechanisms of internal erosion have been recently distinguished [1] (**Figure 1**):

a. *Concentrated leaks*. The cracks are induced by differential settlement during construction or in operation of the embankment, by hydraulic fracture in case of low stresses, compared

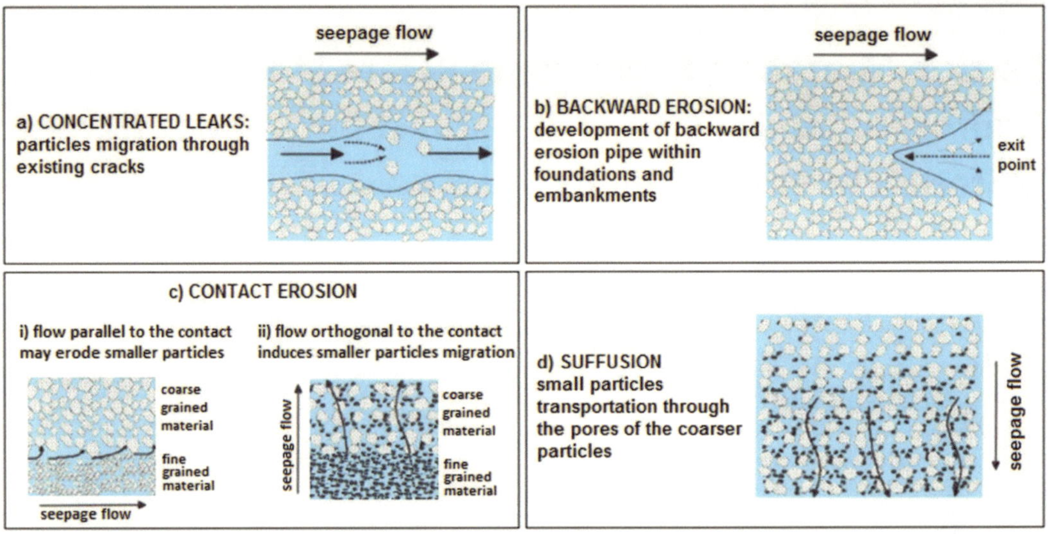

Figure 1. Internal erosion mechanisms.

to the internal pressure, around erosion channels as well as by the desiccation at high levels in the fill. The sides of cracks may be eroded by the leaking water [1].

b. *Backward erosion*. The erosion process begins at a free surface on the downstream side of a dam and evolves beneath or within the embankment, through the development of erosion channels or piping.

c. *Contact erosion*. (i) Seepage flow parallel to the contact between coarse grained materials and fine soils may erode the smaller particles; (ii) particle migration of fine grained materials (base B) through the voids of a coarser materials (transition T), under seepage flow oriented along a direction almost orthogonal to the interface of the contacting materials [2].

d. *Suffusion*. The small particles of soil are dragged by the seepage flow through the pores of the coarser particles.

Concentrated leaks, suffusion, backward erosion (e.g., at the downstream face of the embankment or of the embankment core) may trigger piping erosion phenomena within the embankment or its foundation. The presence of adequate filters that autonomously generate within transition materials can prevent these triggering mechanisms from evolving at a larger scale and forming pipes [3]. Therefore, the *problems of granulometric stability at the contact* between materials characterized by different grain size curves, under seepage forces normally oriented with respect to the average contact surface, assume particular importance. The assessment of transition zones and downstream zones to act as filters is well understood and in many dams it can be demonstrated that these will provide adequate filter protection even if they do not fully meet modern "filter" design criteria [2]. The control of the granulometric stability of base (B) requires a correctly designed protective transition (T) [2, 4]; its voids, related to the grain size distribution and porosity, must be small enough to stop the migrating B particles within short distances, thus avoiding limit states related to the backward erosion possibly appearing under the shape of flow pipes (**Figure 2**); T must also allow a safe drainage of B, avoid clogging and blinding, inducing in turn uncontrolled increase of interstitial pressure.

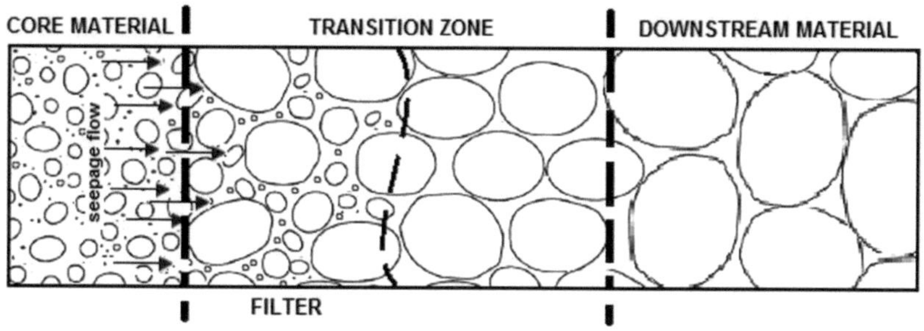

Figure 2. Filter formation mechanism.

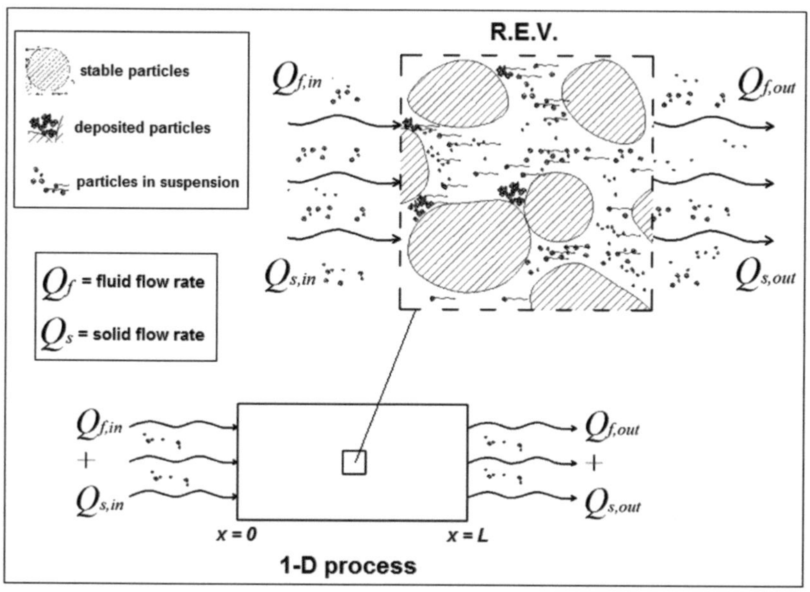

Figure 3. Physical scheme of R.E.V.

2. Particle migration phenomena modeling

By referring to the representative elementary volume (R.E.V.) of a granular material composed by the volume of particles in suspension, the fluid phase volume and the solid phase (*stationary solids*) volume (**Figure 3**), the particle migration process can be described, along space x and time t, by the following system of governing partial differential equations (PDEs):

- *fluid mass balance equation*

$$\nabla \cdot [(1 - c_s) \cdot n \cdot \vec{v}_f] = -\frac{\partial[(1 - c_s) \cdot n]}{\partial t} \tag{1}$$

being $\vec{v}_f$ the velocity vector of the fluid phase;

- *solid mass balance equation*

$$\nabla \cdot (c_s \cdot n \cdot \vec{v}_{sp}) = \frac{\partial[(1 - c_s) \cdot n]}{\partial t} \tag{2}$$

being $\vec{v}_{sp}$ the velocity vector of the transported particles in suspension ($\vec{v}_{sp} = \chi \cdot \vec{v}_f$, with $\chi \in (0; 1]$); c_s, the particles in suspension concentration; n, the volumetric porosity; the contribution of diffusion term has been neglected;

- *"kinetic equation,"* describing the deposition and erosion processes of the deposited/accumulated particles within voids [5]; to this purpose, different formulations have been proposed in the past (**Table 1**) [5–10].

The unknown variables of the problem are n, c_s, v_f, depending on t and x.

In the proposed numerical procedure, the relationship proposed by Indraratna and Vafai (1997) [10] is applied (**Table 1**).

If the following hypotheses are assumed: (i) $v_{sp} = v_f$ (χ =1) and (ii) unidirectional flow (1-D case), since $\vec{v}_D = n \cdot \vec{v}_f$ ($\vec{v}_D$, Darcy's velocity), by combining Eqs. (1) and (2), it is obtained:

$$v_{D,x} \cdot \frac{\partial c_s}{\partial x} = \frac{\partial[(1 - c_s) \cdot n]}{\partial t} \tag{3}$$

For a complete simulation of the particles 1D migration and its evolution towards limit (*clogging, blinding, complete erosion*) granulometric or stability conditions, the (space and time) variability of granulometric properties, voids volume, porosity (n), permeability (k), flow velocity, local piezometric gradients, flow direction, as well as the particles erodibility should be taken into account. However, the variability of the previously defined physical-mechanical variables, especially of the voids volume distribution (VVD) [11, 12], cannot be appropriately described through a simple equation.

Thus, the above formulations may be safely applied to real cases only if the limits deriving from extreme schematization of the analyzed process are removed.

To solve this extremely complex problem, a numerical procedure has been developed. The procedure takes into account grain size curve, constriction sizes distribution (CSD), porosity of the particulate materials, the rate of the suspension, piezometric gradients, drag forces associated with the seepage flow; the friction triggered by normal contact forces induced by confining pressure related to the effective stress state, is considered too.

	Equation	Parameters
Litwiniszyn (1966)	$\frac{\partial n}{\partial t} = c_1[n_0 - n] \cdot c_s - c_2 \cdot n$	• n_0 = initial porosity; • c_1 and c_2 = experimental coefficients (erosion of particles if $c_1 = 0$; $c_2 > 0$; deposition of particles if $c_1 > 0$; $c_2 = 0$).
Sakthivadivel (1966)	$\frac{\partial n}{\partial t} = P_0 \cdot \frac{[n_0 - n]}{n_0} \cdot Q \cdot c_s$	• P_0 = initial probability of deposition of small particles within the voids of the granular material; • Q = fluid flow rate.
Vardoulakis (2004)	$\frac{\partial n}{\partial t} = \Lambda \cdot (1 - n) \cdot c_s \cdot n \cdot v_{sp}$	- Λ = erosion law constitutive (experimental) parameter with dimension of inverse length.
Saada (2005)	$\frac{\partial n}{\partial t} = -a \cdot c_s$	- a = positive scalar
Zhang et al. (2013)	$(1 - c_s)\frac{\partial n}{\partial t} = n\frac{\partial c_s}{\partial t} - k\nabla P_f \nabla c_s + k\rho_f g \nabla c_s$	• k = permeability coefficient, depending on n and c_s; • P_f = fluid pressure; • ρ_f = fluid density.
Indraratna and Vafai (1997)	$\gamma_m = \frac{\gamma_w V_w + \gamma_s V_s}{V_w + V_s} = \frac{(1 - c_s) \cdot \gamma_w + c_s \cdot \gamma_s}{1}$	• γ_w = the unit weight of fluid • γ_s = the unit weight of solid phases; V_w = the fluid volume • V_s = the volume of particles affected by smaller diameter than the average one (d_0) of the voids, defined as $$d_0 = 2.67 \cdot \frac{n}{1 - n} \cdot d_h$$ with $d_h = 1/\sum_i \frac{\Delta P_i}{\bar{d}_i}$; $\Delta P_i = P_{i+1} - P_i$, granulometric passing percentage associated with $\bar{d}_i = (d_{i+1} + d_i)/2$.

Table 1. Some "kinetic" equations available in the literature.

Through the simulation of the 1D coupled particle migration and seepage unsteady states, this procedure aims: (i) to preliminarily evaluate if design protective transition zones, between core and downstream materials (**Figure 2**), are adequate to the "natural" filters formation (time of formation, capability of new-formed filter to limit the transport/erosion of smaller particles); (ii) to understand if, in existing earth dams not provided with transitions, the formation of a "natural" transition zone, due to deposition of smaller particles within the voids of the downstream material (typical of widely graded morainic materials [2]), occurs and allows to interrupt/stop particle erosion or migration.

3. Proposed numerical procedure

To model the particle migration processes and evaluate the safety of earth structures against serviceability or ultimate limit states, an advanced characterization of the granular material as well as the simulation of the (space and time) variability of its properties must worked out.

3.1. Characterization of the granular material

3.1.1. Permeability coefficient

The permeability coefficient (k) represents the fundamental parameter figuring in the equation that describes the seepage of a fluid through a porous medium; it mainly depends on grain size properties and porosity. k can be evaluated through the Kozeny-Carman relationship [13, 14]:

$$k = \chi \cdot \frac{\gamma_w}{\mu_w} \cdot \frac{n^3}{(1-n)^2} \cdot d_h^2 \tag{4}$$

being d_h the equivalent diameter of grains, previously defined (**Table 1**); χ, a numerical coefficient; and μ_w, the water viscosity.

3.1.2. Distribution of the volume of voids and corresponding distribution of constriction sizes

According to a "*geometric-probabilistic*" model [11, 12, 15], the most reliable distribution of the volume of voids within porous media corresponds to a situation of maximum "disorder" of the granular material (particles and voids). The probability that a generic void volume V assumes smaller or equal value than V^* (*cumulative function of probability*) is

$$F(V^*) = \int_{V_{\min}}^{V^*} f(V)dV \tag{5}$$

from which it is possible to write the following "*compatibility*" equations:

$$\int_{V_{\min}}^{V_{\max}} f(V)dV = 1 \tag{6}$$

$$\int_{V_{\min}}^{V_{\max}} V \cdot f(V)dV = \overline{V} \tag{7}$$

being $V_{\min}$ and $V_{\max}$ the minimum and maximum pore volumes, respectively, corresponding to the smallest ($d_{\min}$) and largest ($d_{\max}$) particle diameters ($V_{\min} = 0.16855 \cdot d_{\min}^3$; $V_{\max} = 0.16855 \cdot d_{\max}^3$, if the pore of maximum volume is composed by three particles; $V_{\max} = 0.476 \cdot d_{\max}^3$, if the pore of maximum volume is composed by four particles); $\overline{V}$, the expected value of the pores volume:

$$\overline{V} = \frac{n}{1-n} \cdot \sum_i \frac{\Delta P_i}{\sum_i \frac{6\Delta P_i}{\pi \overline{d}_i^3}} \tag{8}$$

where d_j represents the diameter of a generic particle and ΔP_j its corresponding percentage by weight. By maximizing the configurational entropy associated with the distribution of the pore volumes [12, 15]:

http://dx.doi.org/10.5772/67785

$$E = \kappa' \int_{V_{min}}^{V_{max}} f(V) \cdot \ln[f(V)] dV \tag{9}$$

with $\kappa' = \kappa \cdot N_{vtot}$ (κ is a constant, N_{vtot} is the total number of pores equal to the total number of particles N_{ptot}), through the Lagrange's multipliers method, the following probability density function of pore volumes is obtained:

$$f(V) = \frac{e^{-\beta V}}{\int_{V_{min}}^{V_{max}} e^{-\beta V}} \tag{10}$$

The Lagrange's multiplier coefficient β is numerically deteminated through the following equation, obtained by introducing Eq. (10) in the "*compatibility*" Eq. (7):

$$[\beta \cdot (V_{min} - \overline{V}) + 1] \cdot e^{-\beta V_{min}} - [\beta \cdot (V_{max} - \overline{V}) + 1] \cdot e^{-\beta V_{max}} = 0 \tag{11}$$

Thus, β depends on the porosity n through the expected value $\overline{V}$ of the pores volume.

Through Eq. (10), it is possible to obtain the *cumulative probability function* of the volume of voids:

$$F(V) = \frac{e^{-\beta V} - e^{-\beta V_{min}}}{e^{-\beta V_{max}} - e^{-\beta V_{min}}} \tag{12}$$

For an assigned volume V of a pore, the volume of the largest particle (V_{cs}), able to move through the porous material, satisfies the relation $V_{cs} < V$. By assuming spherical particles (D, diameter), on the basis of geometric observations [11, 15], it is possible to determine the diameters of the smallest ($D_{cs,min}$) and largest ($D_{cs,max}$) particles passing through the smallest ($V_{cs,min}$) and largest ($V_{cs,max}$) pores (**Figure 4**); in other words, the minimum and maximum constriction sizes:

$$D_{cs,\,min} = 2 \cdot (3^{0.5}/3 - 1/2) \cdot D \rightarrow V_{cs,\,min} = 1.94 \cdot 10^{-3} \cdot D^3 \tag{13}$$

(*pores formed by three spherical particles*)

$$D_{cs,\,max} = (2^{0.5} - 1) \cdot D \rightarrow V_{cs,\,max} = 3.72 \cdot 10^{-2} \cdot D^3 \tag{14}$$

(*pores formed by four spherical particles*)

By defining the coefficient $\eta = V_{cs}/V$, it is obtained:

$$\eta_{min} = \frac{V_{cs,\,min}}{V_{min}} = 1.15 \cdot 10^{-2} \tag{15}$$

$$\eta_{max} = \frac{V_{cs,\,max}}{V_{max}} = 7.81 \cdot 10^{-2} \tag{16}$$

Thus, the volume of constriction sizes and the corresponding diameter (CSD) can be generally evaluated as:

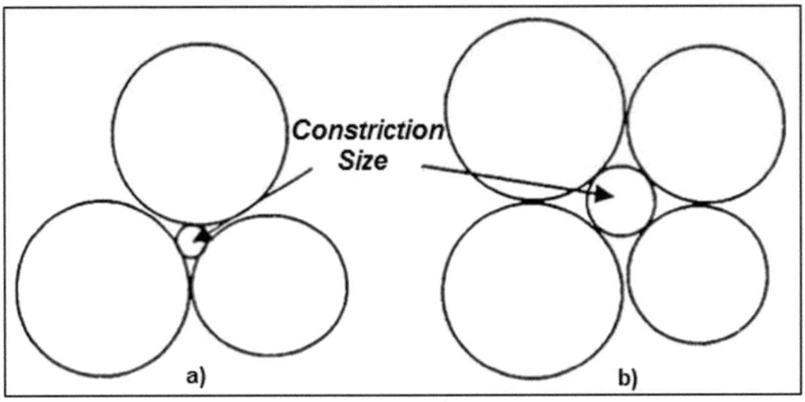

Figure 4. Constriction size for pores formed by (a) three and (b) four particles.

$$V_{cs} = \eta(V) \cdot V \tag{17}$$

$$D_{cs} = \sqrt[3]{\frac{\eta(V) \cdot V \cdot 6}{\pi}} \tag{18}$$

If a linear change of η with volume V is simply assumed, it is obtained:

$$\eta(V) = \frac{1}{V_{\max} - V_{\min}} \cdot [(\eta_{\max} - \eta_{\min}) \cdot V + (\eta_{\min} V_{\max} - \eta_{\max} V_{\min})] \tag{19}$$

Several methods to determine the CSD are available in technical literature [16–22]. Some of them take into account the loose (N_p = 4, being N_p the number of particles forming voids) and dense (N_p = 3) soil states or relative densities [16–21]; other approaches enable the consideration of the porosity (n) resulting in a density dependent CSD [22, 23].

The comparison between some of methods available in literature and the method proposed and adopted by the author, previously described, is carried out (**Figures 5** and **6**).

It is observed that the author's method provides CSD curves ranged between the ones obtained through the models proposed by the authors of Refs. [16, 17, 19], corresponding to the dense soil state (**Figure 5**), and by To et al. [22], for assigned values of porosity n (**Figure 6**).

The porosity n, neglected by the authors of Refs. [16, 17, 19], may greatly influence the CSD; as a result, there might be theoretically the possibility of particle movements under seepage forces even for fairly poorly graded soils if the porosity would be high enough [22].

The comparison between the author's method and well-known experimental results [23] is reported in **Figure 7**. In the considered lab experiment [23], the constrictions sizes of a sample of gravel with grain sizes between 8 and 63 mm and porosity n = 0.34 were manually measured.

http://dx.doi.org/10.5772/67785

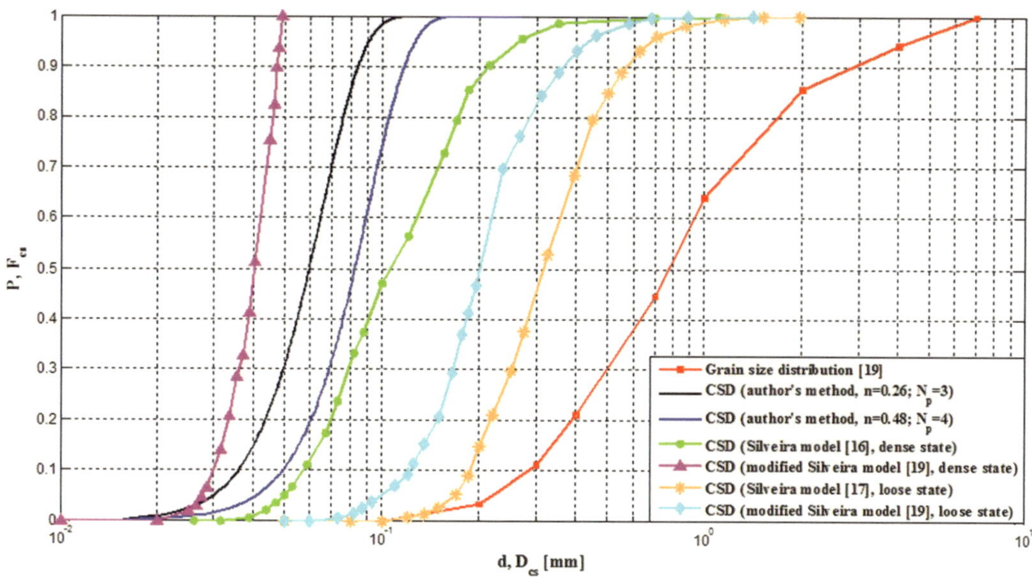

Figure 5. Grain size curve [19] and CSD curves obtained by different methods (*d* and *P* are the diameter and weight percentage related to grain size distribution, respectively; D_{cs} and F_{cs} are the diameter and weight percentage related to CSDs).

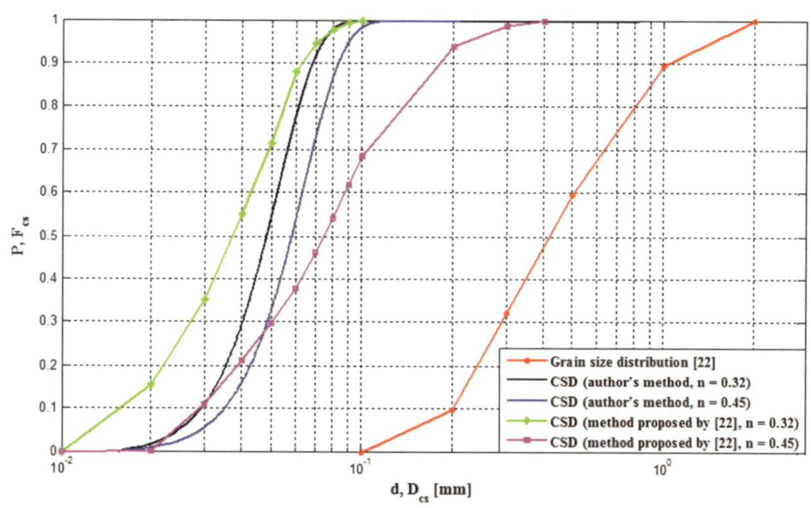

Figure 6. Grain size curve [22] and CSD curves obtained by different methods (*d* and *P* are the diameter and weight percentage related to grain size distribution, respectively; D_{cs} and F_{cs} are the diameter and weight percentage related to CSDs).

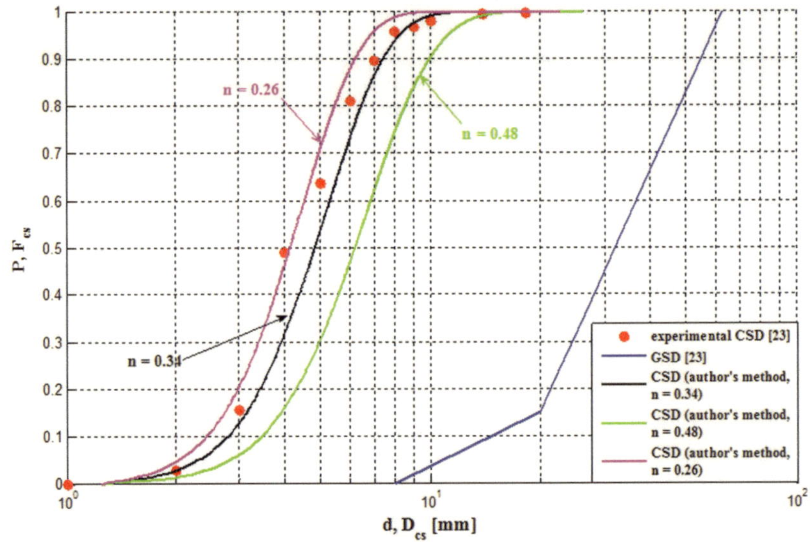

Figure 7. Grain size distribution (GSD) [23], experimental CSD curve [23] and CSD curves obtained by the proposed method (d and P are the diameter and weight percentage related to grain size distribution, respectively; D_{cs} and F_{cs} are the diameter and weight percentage related to CSDs).

It is worth observing that the best interpretation of experimental CSD, through the model proposed by the author, is obtained for the porosity value $n = 0.34$ characterizing the tested material.

3.2. Problem's setting and governing equations

The heterogeneous porous material is decomposed into several elements (**Figure 8a**), each characterized by initial grain size curve ($P_{i,j,0}$), porosity ($n_{i,0}$) and permeability ($k_{i,0}$); i, j, and t define the system element, materials granular fractions and the elapsed time, respectively [24]. Each element is schematically composed by original material ($V_{or,i,t}$), deposited/accumulated particles ($V_{acc,i,t}$; $V_{acc,i,0} = 0$), due to migration phenomena, and water saturating the ith element ($V_{w,i,t}$). According to the Kozeny-Carman Eq. (4), the permeability $k_{i,t}$ is expressed as

$$k_{i,t} = \chi \cdot \frac{\gamma_w}{\mu_w} \cdot \frac{n_{i,t}^3}{(1 - n_{i,t})^2} \cdot d_{h_{i,t}}^2 \tag{20}$$

The variables $P_{i,j,t}$ and $n_{i,t}$ (and then $k_{i,t}$) evolve because of erosion-deposition processes, associated with particle migration, causing an unsteady seepage flow. The unsteady state is simply analyzed by considering a sequence of steady states (time interval, Δt; "successive steady states" method); for each Δt, the continuity equation holds:

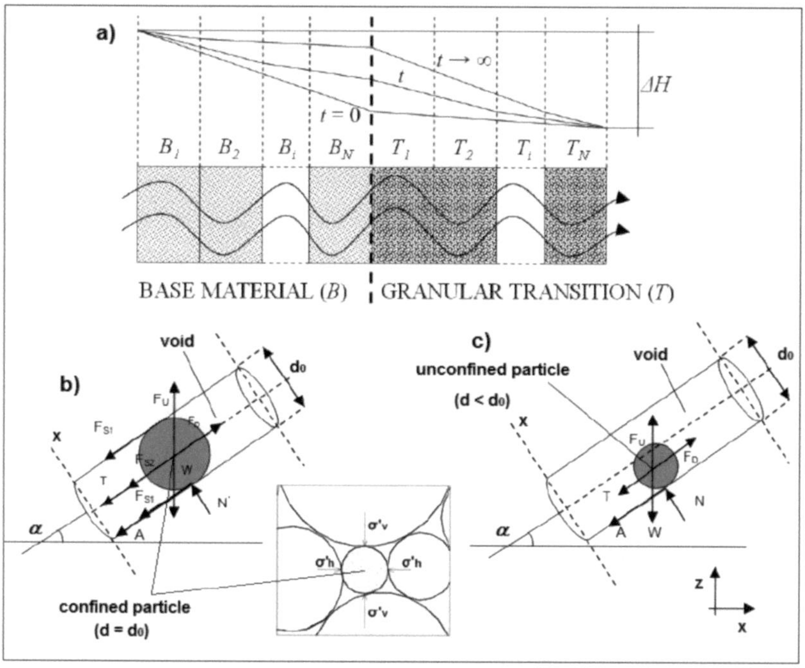

Figure 8. (a) Problem setting. One-dimensional unsteady seepage flow through a heterogeneous base (B)-transition (T) system; B and T are divided into elements; a constant head difference ΔH is imposed. Forces acting on a migrating particle: (b) plugged particle (d_0); and (c) unplugged particle ($d < d_0$); d, particle diameter; d_0, average size of a pore channel (adapted from Ref. [2]).

$$\begin{bmatrix} \frac{k_{1,t}}{l_1} & -\frac{k_{2,t}}{l_2} & 0 & \cdots & \cdots & 0 \\ 0 & \frac{k_{2,t}}{l_2} & -\frac{k_{3,t}}{l_3} & 0 & \cdots & 0 \\ \vdots & \vdots & \vdots & \vdots & \vdots & \vdots \\ 0 & \cdots & \frac{k_{i,t}}{l_i} & -\frac{k_{i+1,t}}{l_{i+1}} & \cdots & 0 \\ \vdots & \vdots & \vdots & \vdots & \vdots & \vdots \\ 0 & \cdots & \cdots & 0 & \frac{k_{N-1,t}}{l_{N-1}} & -\frac{k_{N,t}}{l_N} \\ 1 & 1 & \cdots & \cdots & \cdots & 1 \end{bmatrix} \cdot \begin{pmatrix} \Delta h_{1,t} \\ \Delta h_{1,t} \\ \vdots \\ \Delta h_{2,t} \\ \vdots \\ \Delta h_{N,t} \end{pmatrix} = \begin{pmatrix} 0 \\ 0 \\ \vdots \\ 0 \\ \vdots \\ \Delta H \end{pmatrix} \tag{21}$$

Therefore, the suspension rate Q_t through the elements of the section Ω, and the volume of the suspension $V_{\text{out},i}$, composed by the scoured particles dragged by the seeping fluid, $V_{s,\text{out},i,t}$, entered and washed out $V_{w,\text{out},i,t}$ from each element, is the same during each Δt:

$$Q_t = \Omega \cdot k_{i,t} \cdot \frac{\Delta h_{i,t}}{l_i}; \; V_{\text{out},t} = Q_t \cdot \Delta t \tag{22}$$

Futhemore, $Q_{f,\text{in}} = Q_t$ and $Q_{s,\text{in}} = 0$ are assumed: in the first element of the system, during each Δt, the incoming fluid volume $V_{w,\text{in},t} = V_{s,\text{out},1,t}$. Eq. (3) is thus discretized as

$$v_{D,t}\frac{c_{s,i+1,t}-c_{s,i,t}}{l_i}=\frac{\phi_{f,i,t+1}-\phi_{f,i,t}}{\Delta t} \tag{23}$$

with $\varphi_f = (1 - c_s)\cdot n$. Eq. (23) can be rewritten as

$$v_{D,t}\cdot\Omega\cdot\Delta t\cdot(c_{s,i+1,t}-c_{s,i,t})=\Omega\cdot l_i\cdot(\phi_{f,i,t+1}-\phi_{f,i,t}) \tag{24}$$

Then, it is possible to evaluate the temporal variation of the fluid volume within the elements which constitute the porous material:

$$Q_t\cdot\Delta t\cdot(c_{s,i+1,t}-c_{s,i,t})=(V_{f,i,t+1}-V_{f,i,t}) \tag{25}$$

The particles can be scoured if subjected to a flow velocity greater than the local, critical flow rate (v_{cr}) [14, 25]; the analysis of the actions on a movable particle and the dynamic equilibrium along the flow direction [2] allow the estimate of $v_{cr.}$

In some cases, the drag force F_D (Stokes law) may overcome the maximum local shear force related to the effective weight of the particle and the acting confining stresses (F_S) (**Figure 8b** and **c**); therefore, the particle can be eroded : α is a coefficient allowing to consider the density of the granular matrix ($0 < \alpha \le 4/\pi$); $\alpha = 4/\pi$ for granular matrix composed by spherical particles arranged in hexagonal configuration, most dense state [14, 26].

For a horizontal flow path, v_{cr} is expressed as follows:

$$v_{cr}=\frac{n}{3\mu_w}\cdot\left[(\gamma_s-\gamma_w)\frac{d^2}{6}+\frac{\alpha d}{2}(\sigma_z'+\sigma_y')\right]\tan\varphi \tag{26}$$

(laminar flow)

$$\begin{aligned}\left[\frac{24\mu_w}{\rho_w d(v_{cr}/n)}+\frac{5.6\sqrt{\mu_w}}{\sqrt{\rho_w d(v_{cr}/n)}}+0.25\right]\frac{1}{4}\rho_w\left(\frac{v_{cr}}{n}\right)^2\\-(\gamma_s-\gamma_w)\frac{d}{3}(\sin\alpha+\cos\alpha\tan\varphi)-\lambda\,(\sigma_1'+\sigma_2')\,\tan\varphi\end{aligned} \tag{26.bis}$$

(turbulent flow)

Eq. (33.bis) (turbulent flow regime) must be numerically solved. The original material is subjected to strong confining actions (frictional forces, geometric hindrances); high flow velocities are needed to mobilize the plugged particles. Conversely, the accumulated, unplugged particles, may be easily scoured during simulation; the corresponding critical flow velocity assumes small values (Eq. (26)).

The hydraulic conditions allowing the migration of movable particles are first considered; the analysis of the geometric conditions follows, as long as the previous ones is verified ($v > v_{cr}$). To determine the scoured particles (diameter d_j) composing the (*i*th) element, d_j is compared with the constriction sizes of the *(i + 1)*th element. In particular, for each d_j, the percentage $F_{cs,i,j,t}$ of smaller constriction sizes of the (i + 1)th element is defined (**Figure 9**); if the percentage of

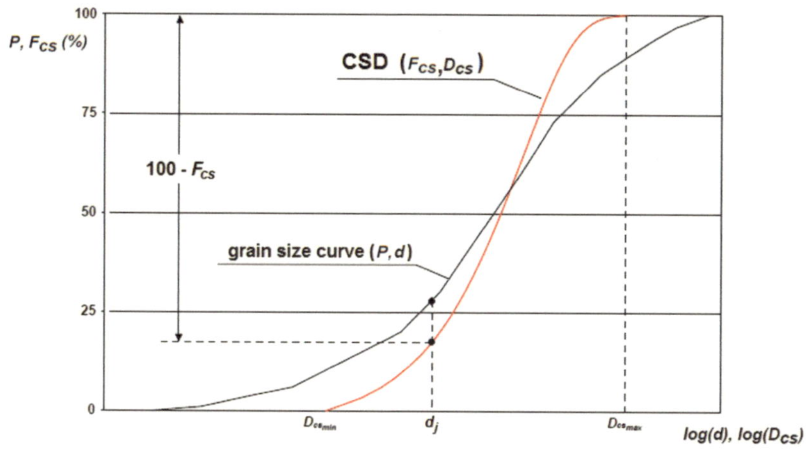

Figure 9. Comparison between particles (diameter d_j) and CSD.

openings through which particles (d_j) may pass ($100 - F_{cs,i,j,t}$) is null, the considered (d_j) and higher remaining granulometric fractions are not able to cross through the voids of the (i+1)th element and then deposit; conversely, the smaller granulometric fractions can be scoured (if $v > v_{cr}$).

The balance of the eroded and deposited granular fractions allows to redefine the grain size distribution, the porosity and the permeability of each element and, then, to determine the piezometric gradients and flow velocity associated with each element of system. Once the fractions of accumulated ($S_{\text{acc},i,t}$) and original eroded ($S_{\text{or},i,t}$) material are determinated through hydraulic and geometric methods, it is possible to evaluate the specific weight of filtering suspension, by rearranging the equation proposed by Indraratna and Vafai [10]:

$$\gamma_{m_{i,t}} = \frac{\left(\frac{S_{acc_{i,t}}}{100} V_{\text{acc}i,t} + \frac{S_{\text{or}_{i,t}}}{100} V_{\text{ori},t}\right)\gamma_{s_i} + V_{w_{i,t}} \cdot \gamma_w}{\frac{S_{\text{acc}_{i,t}}}{100} V_{\text{acc}i,t} + \frac{S_{\text{or}_{i,t}}}{100} V_{\text{ori},t} + V_{w_{i,t}}} \tag{27}$$

The material scoured from the ith element, at time t ($V_{s,\text{out}}$), depends on the specific weight of the filtering suspension, γ_m [10]:

$$V_{s,\text{out}_{t,i}} = \frac{\gamma_{m_{t,i}} - \gamma_w}{\gamma_s - \gamma_w} \cdot V_{\text{out},t} \tag{28}$$

The scoured volume of material is composed both by $V_{\text{or,out}}$ and $V_{\text{acc,out}}$ [27]:

$$V_{s,\text{out}_{t,i}} = V_{\text{or, out}_{t,i}} + V_{\text{acc, out}_{t,i}};\ V_{\text{acc, out}_{t,i}} = \frac{V_{\text{acc}_{t,i}} \cdot S_{\text{acc}_{i,t}}}{V_{\text{acc}_{t,i}} \cdot S_{\text{acc}_{i,t}} + V_{\text{or}_{t,i}} \cdot S_{\text{or}_{t,i}}} \tag{29}$$

$V_{or,\text{out}}$ and $V_{acc,\text{out}}$ are decomposed into their granular fractions [10]:

$$\Delta V_{\text{or, out}_{i,t,j}} = V_{\text{or, out}_{t,i}} \cdot \frac{P_{\text{or}_{i,t,j}} - P_{\text{or}_{i,t,j-1}}}{S_{\text{or}_{i,t}}};\ \Delta V_{\text{acc, out}_{i,t,j}} = V_{\text{acc, out}_{t,i}} \cdot \frac{P_{\text{acc}_{i,t,j}} - P_{\text{acc}_{i,t,j-1}}}{S_{\text{acc}_{i,t}}} \quad (30)$$

The element within which each scoured fraction is deposited may be determined by taking into account the corresponding length $L_{\text{mig},j}$ of the migration path; $L_{\text{mig},j}$ depends on the probability of a particle not encountering a smaller constriction size. The length covered by an assigned particle up to its arrest is based on concepts of stereology; it depends on the constriction sizes distribution (CSD) related to the PSD (pores size distribution) as well as on the thickness of the filter [12, 15]. The length of the migration path is then compared to the length that the particles can cross during each step Δt: $L_{\text{mig},j} = \min(s \cdot m_j; U_t \cdot \Delta t)$, m_j being the number of constrictions greater than the particle size encountered by the particle along its path; s is the unit step assigned to each comparison (**Figure 10**). At time $t + \Delta t$, the accumulated and original volume fractions within each element become [28]:

$$\Delta V_{\text{acc}i,t+1,j} = \Delta V_{\text{acc}i,t,j} + \Delta V_{\text{accin}i,t,j} - \Delta V_{\text{accout}i,t,j};\ \Delta V_{\text{or}i,t+1,j} = \Delta V_{\text{or}_{i,t,j}} - \Delta V_{\text{or, out}_{i,t,j}} \quad (31)$$

3.3. Validation

Theoretical analyses have been carried out to validate the proposed numerical procedure by simulating laboratory tests carried out by different authors on selected materials [2, 24].

3.3.1. *Results by Atmazidis*

Atmazidis [28] analyzed the particle migration phenomena at the contact between sands (base, B) and gravels (transition, T), under horizontal seepage flow and constant hydraulic gradient (i = 0.25) conditions. The amount of sand deposited within the gravel pores was measured after each test.

The main features of the experiments and the properties of the tested materials (e.g., grain size curves) are shown in **Table 2** and **Figure 11**.

Figure 12 shows the sand content retained in each gravel T material, migrated from the B material, at the end of each test, in function of the distance from sand-gravel interface.

Through the proposed numerical procedure, the experimental results [28] have been back analyzed. The following input parameters have been considered: T material (gravel) thickness = 2.5 m (experimental value); number of elements in which T material has been divided = 20 (each 12.5 cm length); number of elements in which B material has been divided = 1 (1 m length). The B material erosion resistance has been neglected [2].

The results, in terms of particles migration distance, obtained through the proposed procedure, appreciably approximate the values measured by Atmazidis [28]. Particularly, for "clean" gravel (Gravel 1, 0% initial sand content), particles migration distances are larger than those ones obtained for gravels initially containing small sand content (Gravel 2 and Gravel 3) (**Figure 12**).

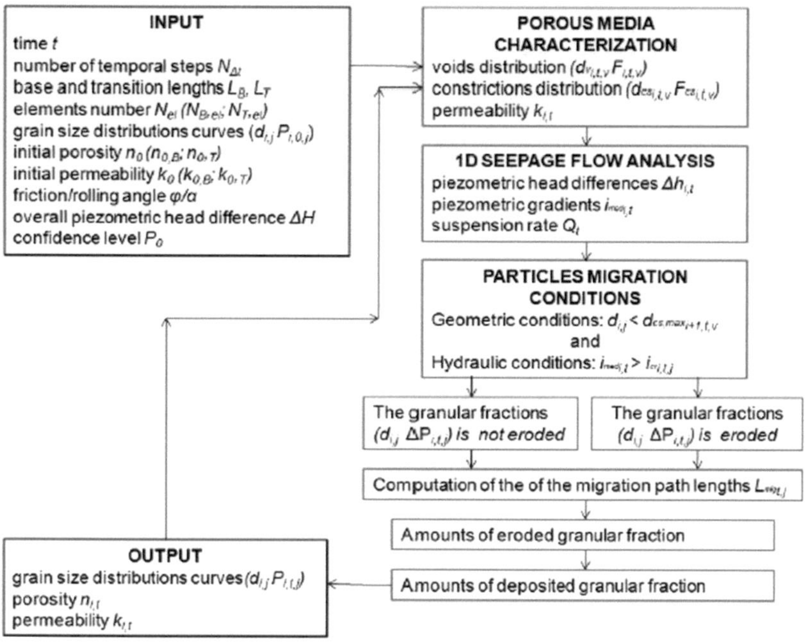

Figure 10. Proposed numerical procedure flowchart.

Test number	5	6	7	8
Grain size curves of coupled sand and gravel	A,3	A,2	B,1	A,1
Sand content of gravel before test (% by weight)	8%	4%	0%	0%
Maximum sand migration length (cm)	75	110	125	215
Amount of sand deposited within gravels (kg) at the end of test	3.17	7.88	9.25	14.5
Negligible migration after (hours)	3.25	3.50	3.00	6.50

Table 2. Synopsis of Atmazidis experimental tests (adapted from Ref. [28]).

Furthermore, the numerical procedure allows to simulate the development of a filter at the sand (B)-gravel (T) interface, able to contrast the sand particle migration: high values of the sand content in the upstream layer of the gravel (15 ÷ 25% by weight, **Figure 12**) mean that a filter is formed [2].

3.3.2. Results by Skempton and Brogan

The effects of an upward seepage flow through a "gap graded" material, under increasing flow rate (Q) conditions, were measured [29]. The progressive increase of Q was obtained by increasing the hydraulic head difference between the lower and the upper surface of the material sample, until the piping process was triggered.

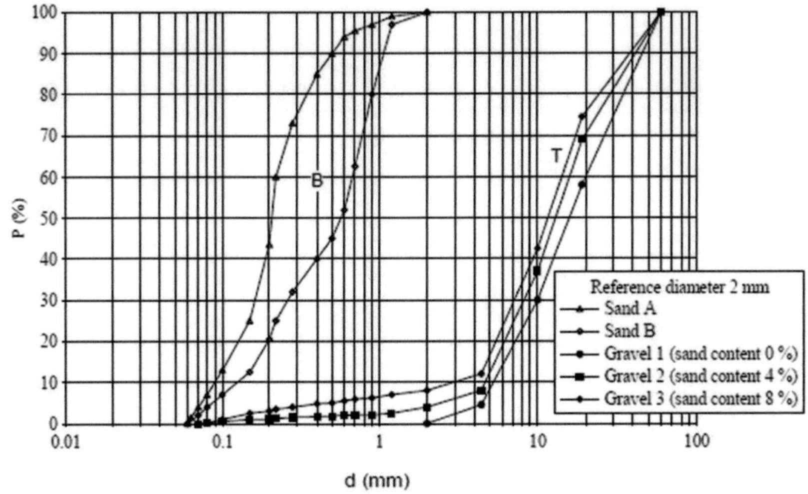

Figure 11. Grains size curves of involved materials (transition, T; base, B), adapted from Ref. [2].

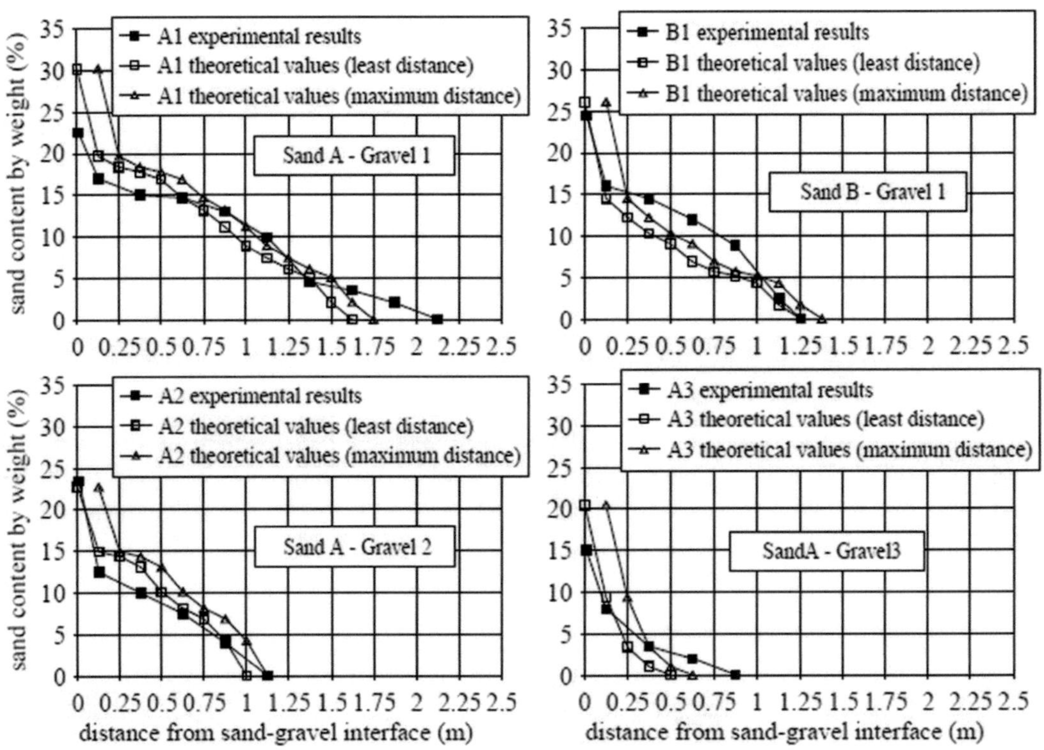

Figure 12. Sand content (s.c.) retained in the gravel transition at the end of the filtering process; theoretical vs experimental results: (a) s.c. after 6.5 hours (Sand A-Gravel 1); (b) s.c. after 3 hours (Sand B-Gravel 1); (c) s.c. after 3.5 hours (Sand A-Gravel 2); and (d) s.c. after 3.25 h (Sand A-Gravel 3), adapted from Ref. [2].

Sample	*k* (m/s)	*N*	Gravel (%)	Sand (%)
A	0.0045	0.34	85 (G)	15 (S1)
B	0.0084	0.37	85 (G)	15 (S1)

Table 3. Main properties of tested materials.

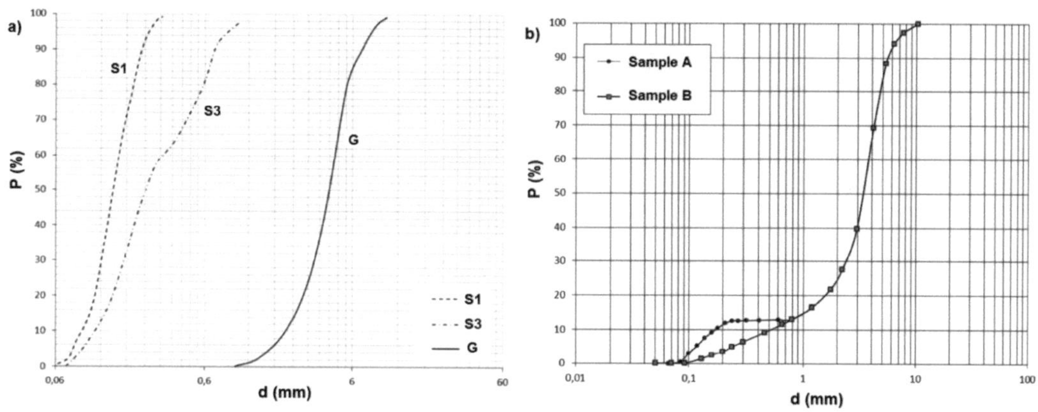

Figure 13. Grain size curves: (a) materials used to prepare the samples and (b) tested samples.

The main properties and the grain size curves of the considered tested material samples are reported in **Table 3** and **Figure 13**, respectively [29].

In all experiments, it is observed an initial linear increase of the average seepage rate (q), following the rise of the average hydraulic gradient (i), until a critical value is achieved; afterwards, the ratio q/i progressively grows if i increases; the erosion of fine particles occurs; while, coarse (gravel) particles are not scoured [24].

To interpret the lab results, each sample (length $L = 0.16$ m) has been divided in eight elements (length 0.02 m); piezometric head $\Delta H = i^*L$ (with $i \in (0, 0.28)$ and $\in (0, 0.37)$ for sample A and B, respectively) has been imposed. Under laminar flow condition, the numerical results seem to well interpret the experimental ones, especially for $i < 0.10$ (sample A) and $i < 0.17$ (sample B); for higher values of i, the numerical values follow a linear trend, underestimating the experimental ones. Under turbulent flow condition, for high values of i, the seepage velocity exponentially increase and the numerical results well fit the experimental ones (**Figures 14** and **15**). The difference between two flow regimes (laminar and turbulent) is due to higher drag forces in the turbulent regime inducing particles migration and, consequentially, the increase of seepage velocity.

4. Analysis of the empirical Terzaghi's criteria

Simulations of the erosion process of B material at the interface B-T have been carried out [30] for two B materials (extended (B1) and homogeneous (B2) grain size distribution GSD). For

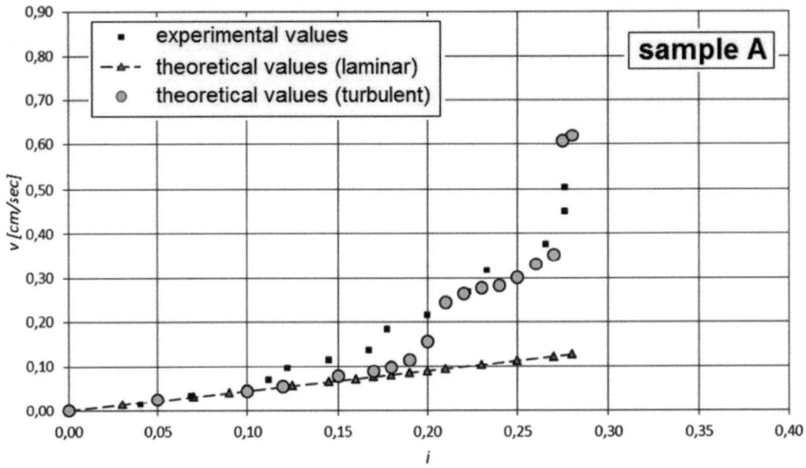

Figure 14. Numerical results vs experimental values: seepage flow (v) vs i, sample A.

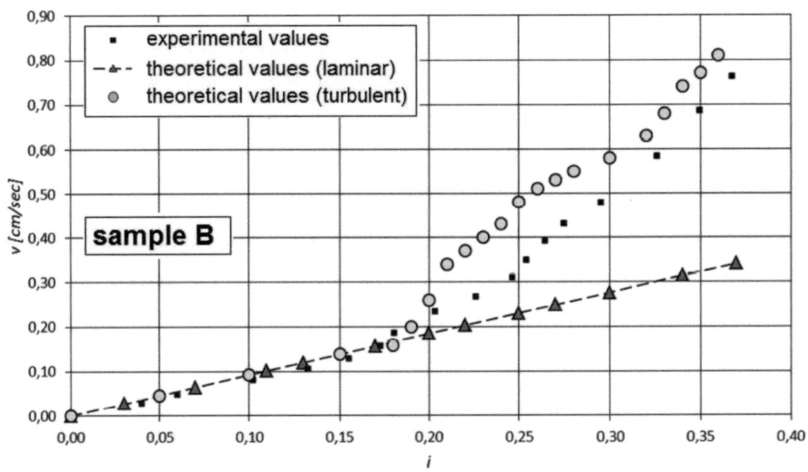

Figure 15. Numerical results vs experimental values: seepage flow (v) vs i, sample B.

each B, different transitions T, matching or not the Terzaghi's criteria [30, 31], have been considered (**Figure 16a** and **b**). Results obtained for some B-T combinations are reported in **Figures 17–21** in terms of evolution of permeability, porosity, average hydraulic gradient, flow velocity, and flow rate. To better understand the phenomenon, results for the combination B1-T3 are reported in **Figures 22** and **23**: the base has been divided into 10 elements; the transition has been divided in 30 elements. At the end of the analysis (t = 120 hours) particles of the eighth, ninth, and 10th element of the base are migrated within the first element of the transition forming the filter.

http://dx.doi.org/10.5772/67785

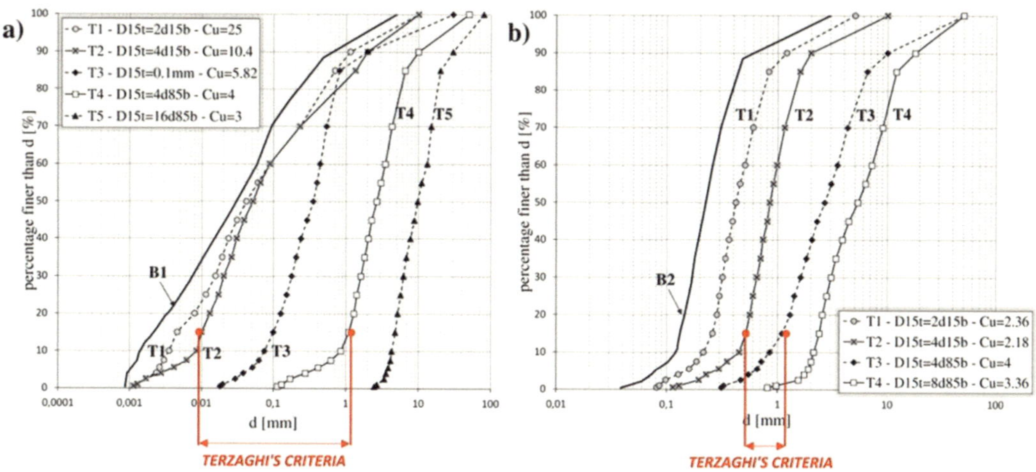

Figure 16. Grain size distributions of bases and transitions analyzed: base B1 (a), base B2 (b). Terzaghi's criteria have been highlighted (piping ratio $D_{15f}/D_{85b} < 4$, permeability ratio $D_{15f}/D_{15b} > 4$), adapted from Ref. [32].

In **Figure 22**, the CSD (constriction size distribution) of the filter at the interface B1-T3 is shown. Filter's voids, related to the GSD of the 10th element of the base, are small enough to stop the migrating particles, according with Terzaghi's criteria. Analysis of the results, moreover, allows to identify areas for grain size distributions of the transitions that show a similar behaviour (**Figure 23a** and **b**). For B1, a first zone which incorporates the *permeability ratio* (finer transitions) is identified: the filter developes rapidly, in reason of the GSD similar to the B material, which minimizes the opening of the voids. The small percentage of fine particles of the T fosters backward erosion, which developes downstream (no confinement) back to the interface B-T. The stabilization of the phenomenon, therefore, is not guaranteed. The second zone has its upper limit in the *piping ratio*: the GSD of the transitions is different from the *B* one. The permeability of *T* is high while hydraulic gradient is small. The weight increase of the matrix particles close to the drain and, as a consequence, the increase of the resistance force opposite to the particles displacemnts strongly reduce the possibility of downstream backward erosion. The B-F-T system quickly reaches a configuration of granulometric equilibrium. A third zone, with coarser *T*, is delimited by a limit T, beyond which it is observed the filter (*F*) is not formated at the interface. Migrating particles, in fact, cross the transition to the drain without being intercepted. For B2, the first zone, with the finer transitions, is located before the *permeability ratio*: the filter quickly forms at the interface; however, backward erosion phenomena downstream are observed because of the small weight of the finer particles of the transition close to the drain (no confinement). The stabilization of the phenomenon is not ensured; for high values of ΔH, a rise of the flow rate is observed. The second zone is defined by the *permeability ratio* as lower limit and the *piping ratio* as upper limit. The influence of the hydraulic load is negligible and no backward erosion downstream is observed. The stabilization of migration is ensured. The third zone has features similar to those identified for the base B1 [32].

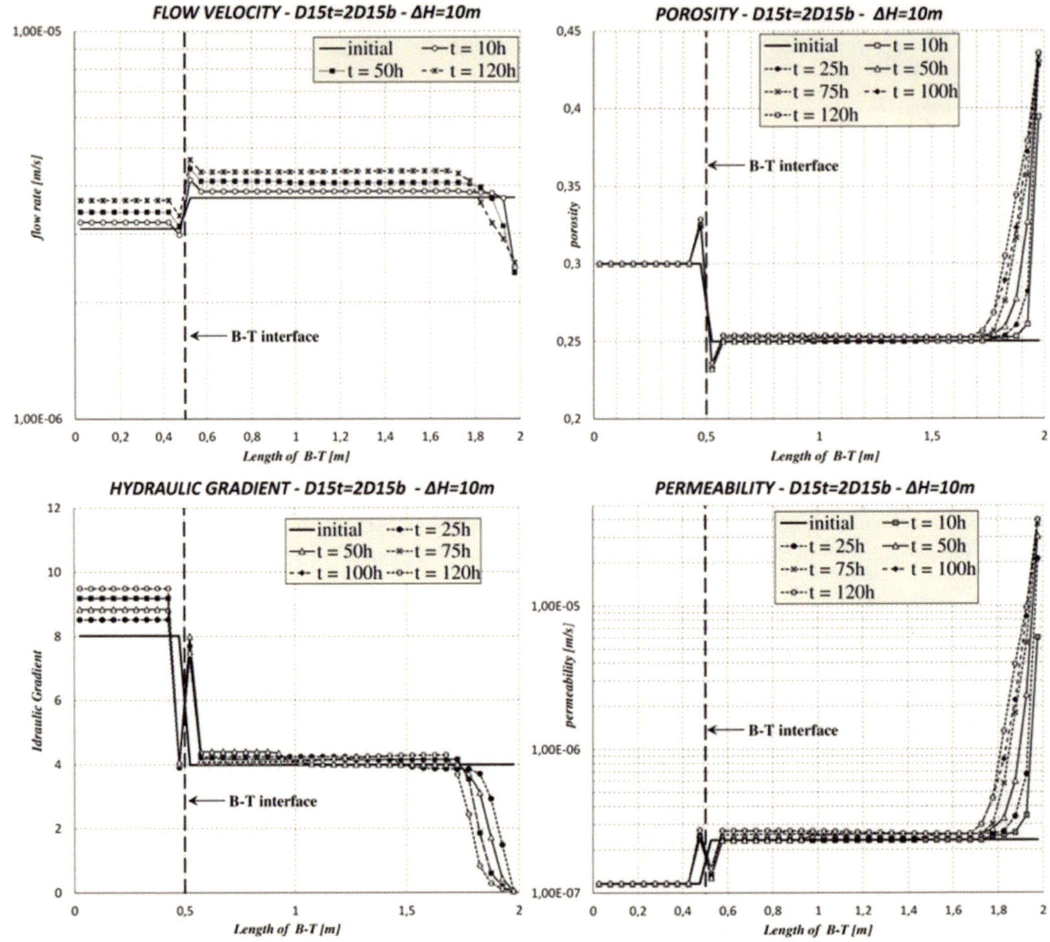

Figure 17. Simulation results for B1-T1 system. It is worth observing a downstream backward erosion; stabilization of the particle migration phenomenon is not attained. Simulations repeated at ΔH = 5 m and ΔH = 20 m showed a remarkable influence of the hydraulic head, adapted from Ref. [32].

5. Analysis of cases

5.1. Suorva dam

Sinkholes and leaks occurred through the moraine core of Suorva dam (Sweden) during periods of floods and high water level in 1983 [33]. In one case, in October 1983, up to 100 l/s of turbid leakage was seen but reduced after 10 days to about half this amount. **Figure 24** shows the dam section and the possible leakage routes. The upper part of the dam was constituted by a coarse material, unable to contrast the internal erosion. The intermediate parts were constituted by materials with seal capacity only after an excessive erosion that may induce damages and sinkholes, as occurred at Suorva. Particularly, the erosion occurred in the core, through the entire width of the filter. This avoided the free downward drainage in the

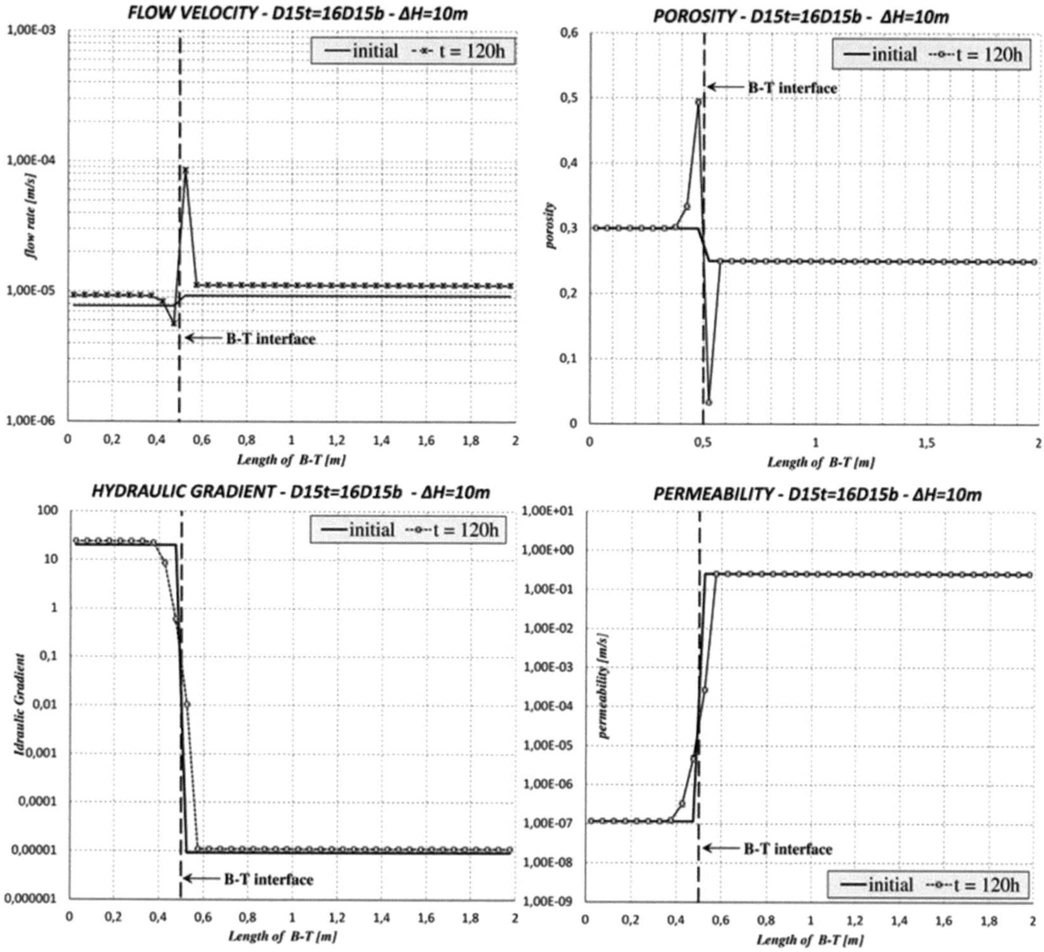

Figure 18. Simulation results for B1-T5 system. Downstream backward erosion is negligible; stabilization of the particle migration phenomenon is reached. Simulations repeated at $\Delta H = 5$ m and $\Delta H = 20$ m showed no remarkable influence of the hydraulic head, adapted from Ref. [32].

filter, maintaining the water level in it to only 2 m below reservoir level for a relatively long time. The dam was not provided with any drain along the filter or into the downstream rockfill. In the lower part of the dam, where there was a fine and a coarse filter, the filter was finer, with a maximum d_{15} of 1.0 mm, compared to a Sherard recommended size of 0.7 mm. Such "some-erosion" filters would soon seal if erosion initiated.

There were indications of erosion at the core-foundation interface, but no extensive damage. The grouting records along the dam at the position where the leaks and sinkholes occurred show large grout takes at the possible leakage positions high in the core and lower in the core; and at the base of the core and into the upper parts of the foundation. To protect the dam against internal erosion and provide it to a drainage system to discharge the leakage flows caused by future incidents, a rockfill berm was built [33] on the downstream slope.

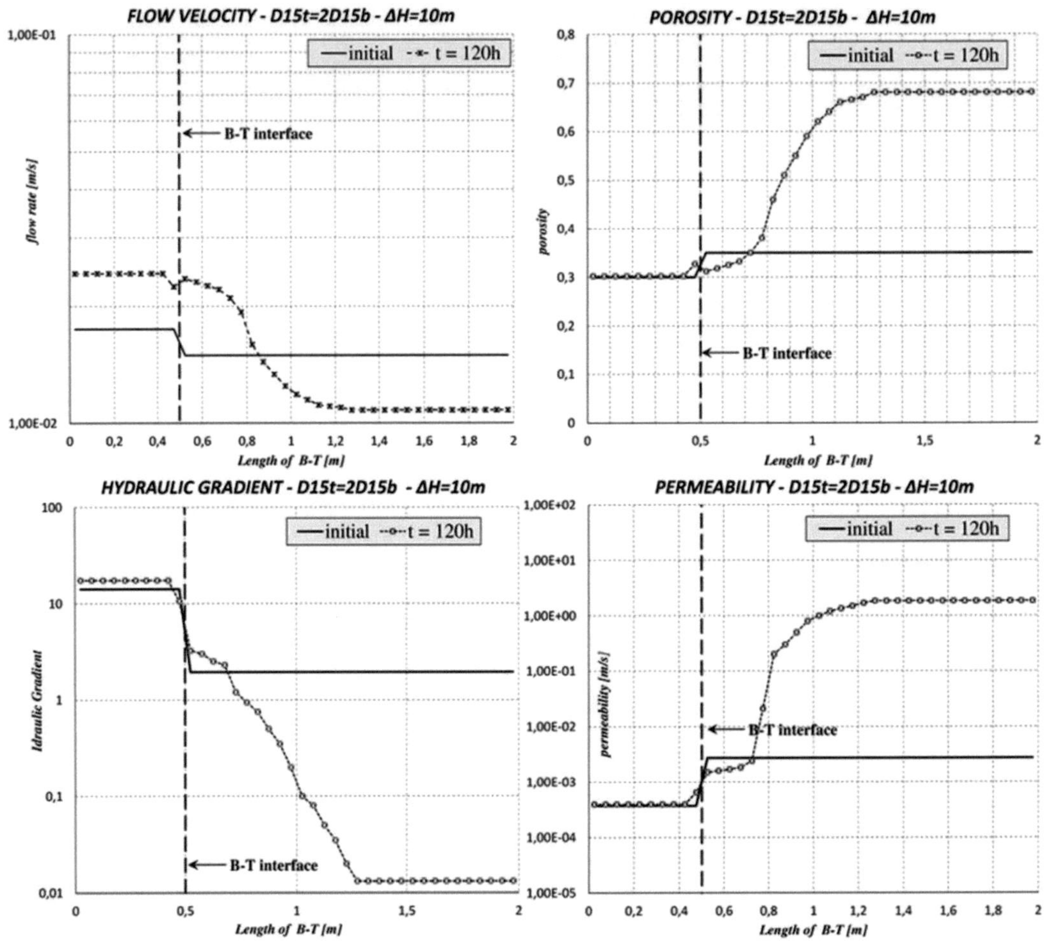

Figure 19. Simulation results for B2-T1 system. The downstream backward erosion is appreciable; stabilization of the particle migration phenomenon is not guaranteed. Simulations repeated at $\Delta H = 5$ m and $\Delta H = 20$ m showed a remarkable influence of the hydraulic head; if ΔH = 20 m, an uncontrolled increase of flow rate can be observed, adapted from Ref. [32].

5.2. San Valentino dam

The 32 m high San Valentino "zoned" embankment dam was built in the 1940s in the basin of Adige river, North Italy, near the small town S.Valentino alla Muta (Bolzano); the dam is 447 m long and 7 m wide; the maximum reservoir elevation is 1498.10 m a.s.l. (**Figure 25a**). The dam project was modified several times to take into account the expected settlements and flow rates; in 1942, an advanced project was elaborated; the construction works were soon interrupted during the Second World War; construction ended in 1950 [34, 35].

Fan-outwash materials (moraine debris with minor amounts of sand and silt) mainly affect the left side of the valley; the subsoil is mainly composed by fine gravel, sand and silt layers, with

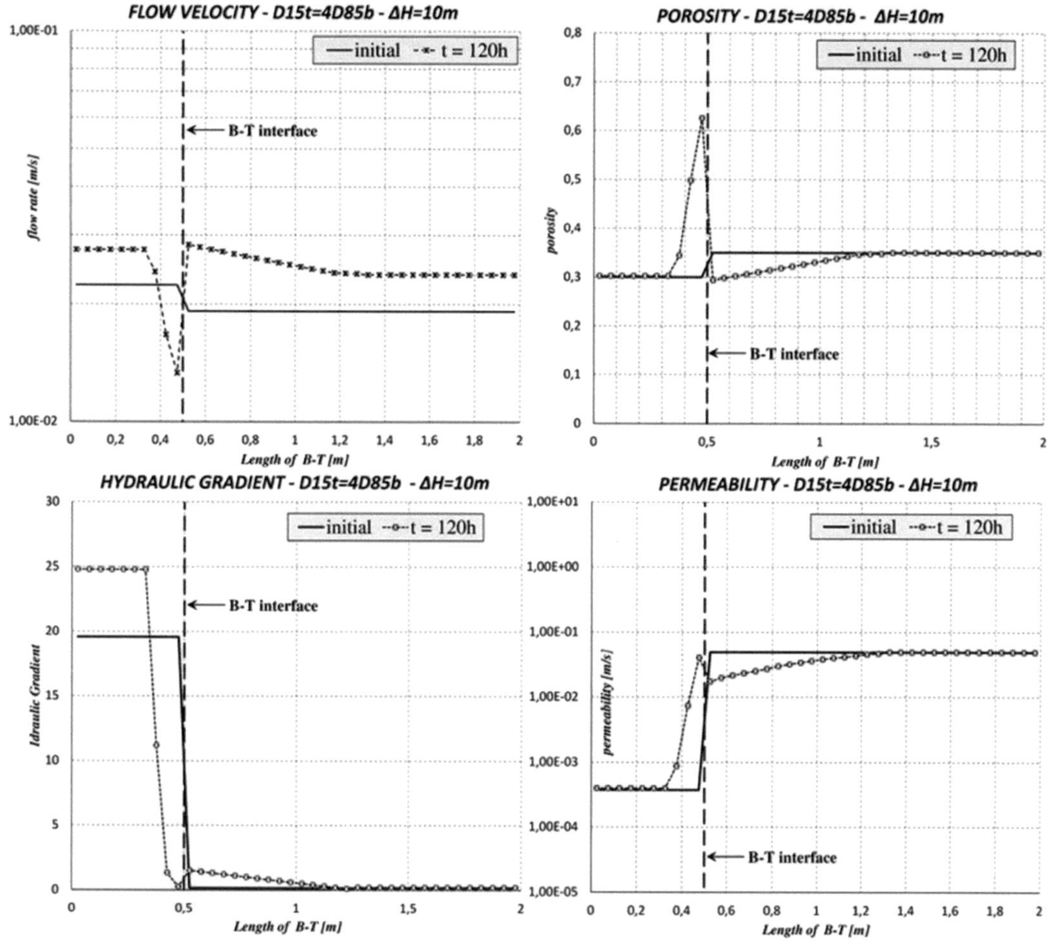

Figure 20. Simulation results for B2-T3 system. Downstream backward erosion is negligible; stabilization of the particle migration phenomenon is not attained. Simulations repeated at $\Delta H = 5$ m and $\Delta H = 20$ m showed no remarkable influence of the hydraulic head. Length of the filter is about 70 cm, adapted from Ref. [32].

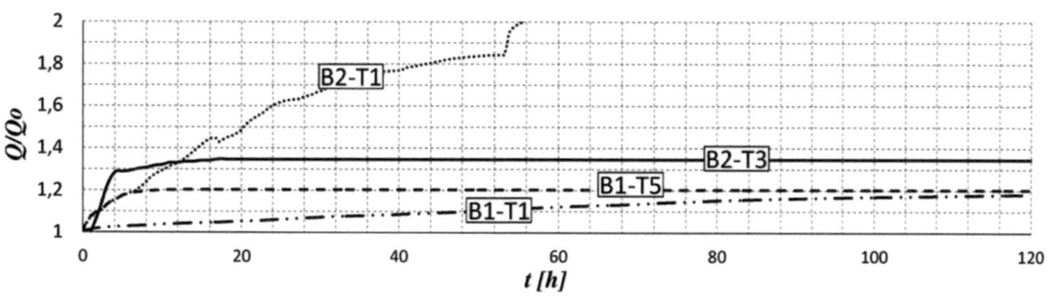

Figure 21. Non-dimensional discharge flow Q/Q_0 along time t for systems B1-T1, B1-T5, B2-T1 e B2-T3. B1-T1 and B2-T1 systems show the continuous increase of the discharge flow along time; for systems B1-T5 and B2-T3, the flow stabilizes after a few hours at ≈20–30% of their initial value, adapted from Ref. [32].

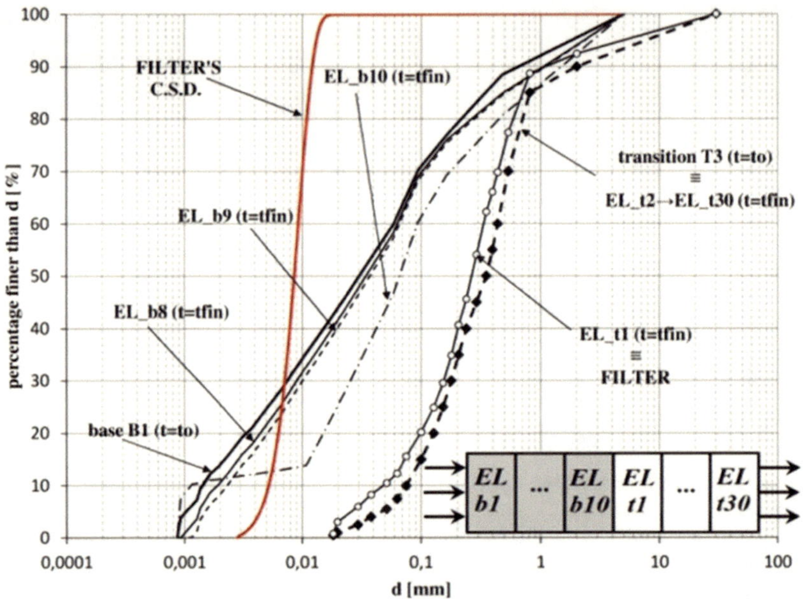

Figure 22. Evolution of grain size distribuition in the elements of B1-T3. The CSD of the filter is reported, adapted from Ref. [32].

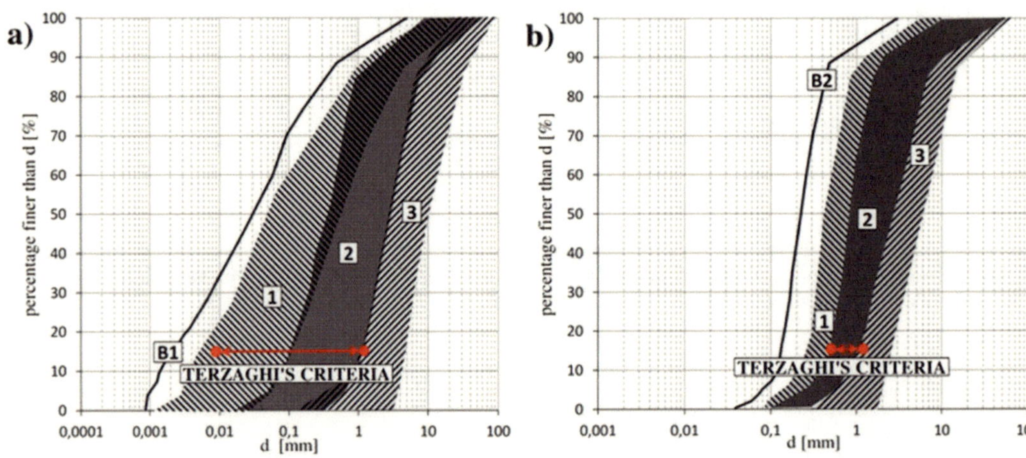

Figure 23. Transitions (T)-base (B) systems, with graded (a) or almost uniform (b) grain size curves, adapted from Ref. [32].

peaty lenses, of alluvial and lacustrine deposition, characterized by significant thickness and low permeability [35].

On the right side, a fissured bedrock outcrops or lies at shallow depth. In the central part, the bedrock dips steeply and at the bottom of the valley it disappears under the sediments mantle

http://dx.doi.org/10.5772/67785

(**Figure 25b**). The fan-outwash material was selected for the dam construction. The grain size of the quarried soils uniformly extends into the range of sand, gravel and silt with negligible amounts of clay. The core grains had d_{max} = 50–60 mm; to reduce the core per-meability, a small amount of bentonite was added, without appreciable effects (**Figure 25c**). The embankment material was compacted at w_{opt} = 6.6% (γ_d = 20.8 kN/m^3).

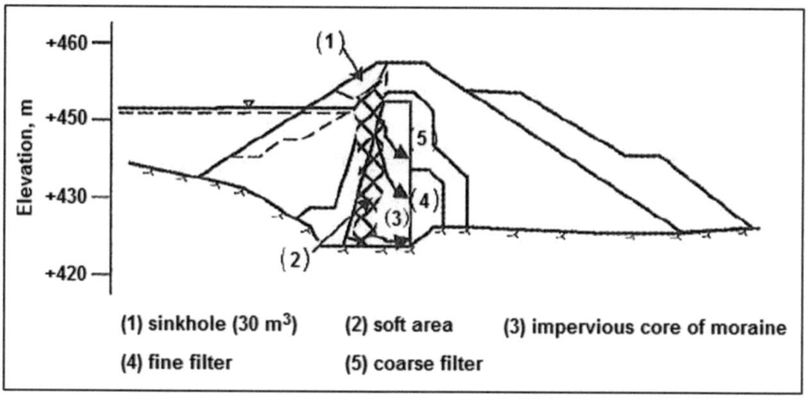

Figure 24. Suorva dam: possible leakage paths during 1983 incident (adapted from Ref. [33]).

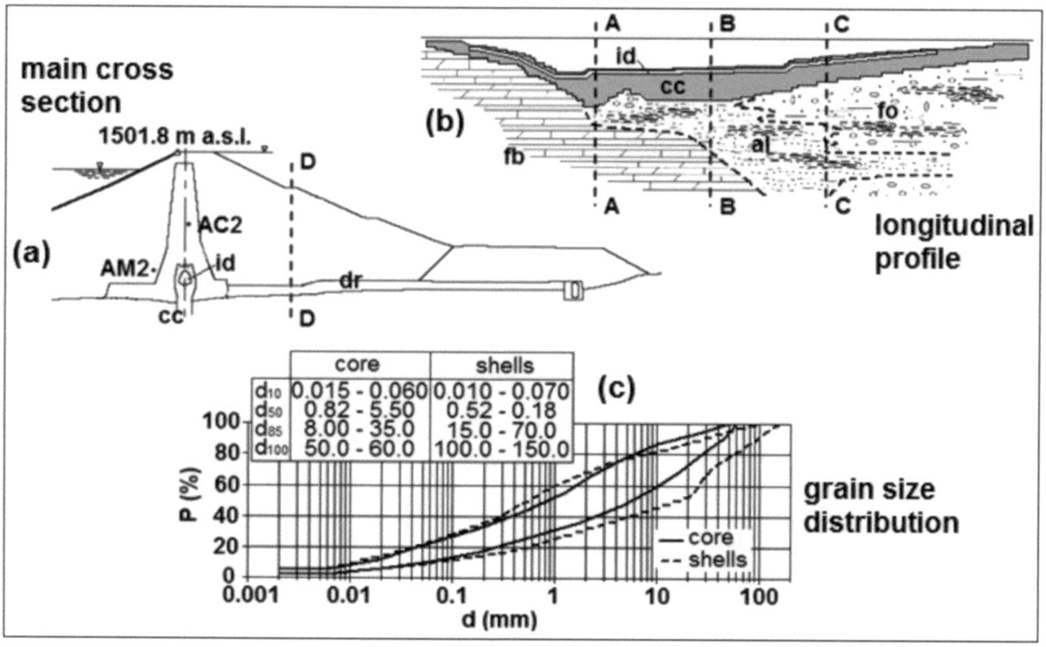

Figure 25. San Valentino dam: (a) main cross section (*A-A*), settlements and piezometric measuring points; (b) longitudinal profile: (*cc*) concrete cutoff (grey colored), (*fb*) fissured bedrock, (*al*) alluvium, (*fo*) fan-outwash; and (c) grain size of materials (adapted from Ref. [34]).

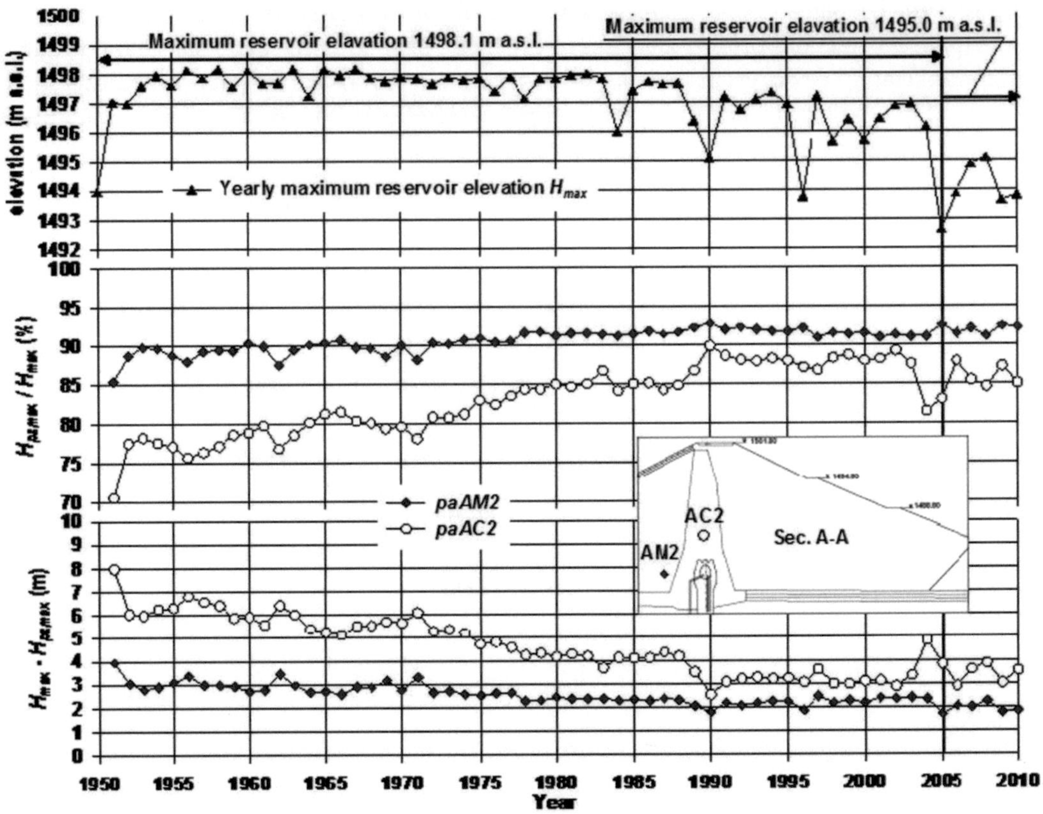

Figure 26. San Valentino dam: Ratio and difference between the yearly maximum pie-zometric heads $H_{pz,max}$ (piezometers *paAM2* and *paAC*, Sec. A-A) and the yearly maximum reservoir elevation H_{max} (the piezometric heads are referred to the elevation of the lower point of the valley 1470 m a.s.l.), adapted from Ref. [34].

Piezometers and assestimeters were installed during construction and operation. In the central part of the dam, small and slow increments of the piezometric heads (p.h.) were measured. In the period 1950–1985, the ratio between the measured maximum value of p.h. and the maximum reservoir level increased about 15% (sect. A-A, **Figure 26**). Afterwards, it became almost constant [35].

Complex FEAs have been deviced to evaluate the properties (geometry, permeability and stiffness heterogeneity) of foundation soils and dam body materials, concrete cutoff, permeability defects of the drain and of the "nominal core", unsaturated hydraulic behavior of materials [34].

Seepage analyses results show that the measured free surface, piezometric heads and leakages cannot been numerically simulated if a homogeneous dam is considered; it is necessary to impose a *k*-zoning of the dam body. Results also show that the current *k* distribution differs from the end of construction *k* distribution (**Figure 27**), due to suffusion phenomena that affected the core material, mainly the finer fractions. The interpretation of the piezometric

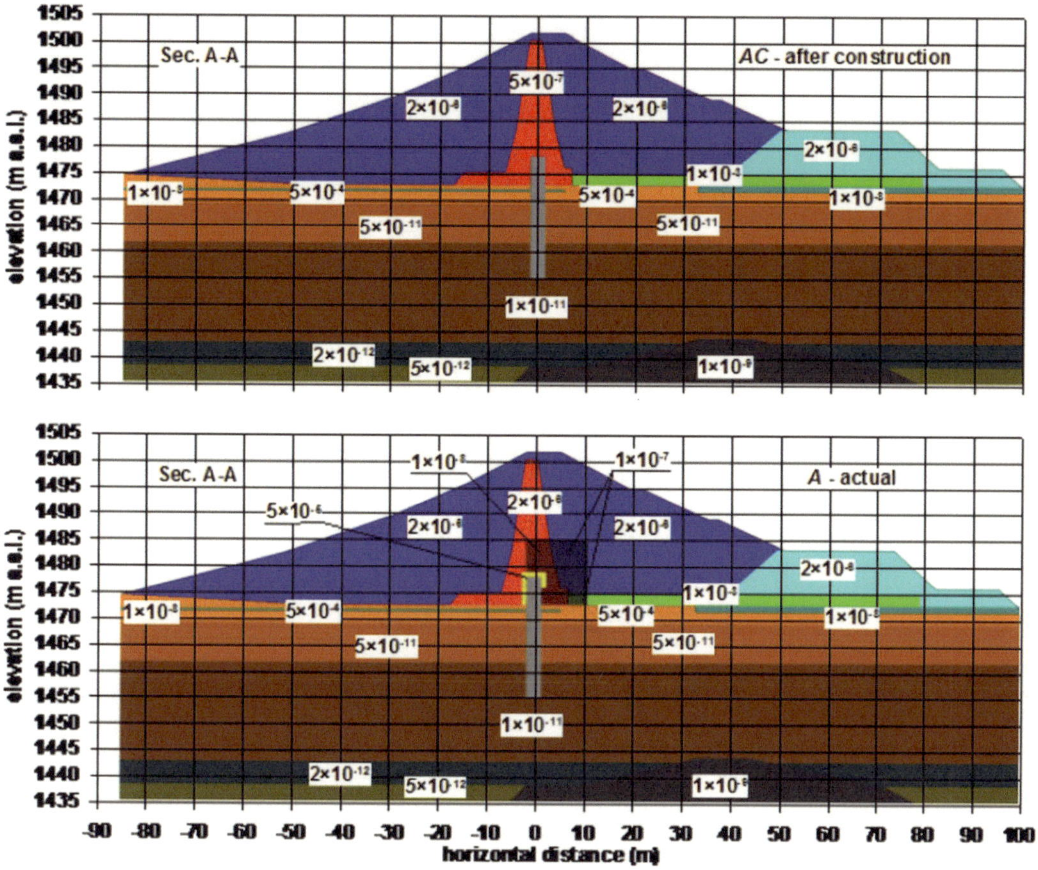

Figure 27. San Valentino dam: permeabilities *k* (m/s) of dam materials and foundation soils, (*AC*) after construction and (*A*) current (adapted from Ref. [34]).

heads measured near the cutoff allows to hypothesize a *local permeability defect* (local increase of *k*) immediately above the diaphragm wall, at the contact with the embankment materials.

According to this hypothesis, properly modeled, an appreciable agreement (both max values and time evolution) among measured and simulated piezometric heads (**Figure 28**) and leakages values, is obtained [34].

5.3. Results

Migration phenomena regarding the moraine materials of the S. Valentino and Suorva dams [2, 27] have been simulated through the proposed numerical procedure (**Table 4**). The following input parameters are assigned: total length of the B-T systems = 3 m (for B, 1 m; for T, 2 m); number of elements dividing the B-T system = 60 (each 5 cm length); piezometric head difference H = 6 m [27]. To favor their migration, B material particles are considered unplugged (i.e., confining stresses are neglected). Numerical results point out that the finer fractions ($d \approx 0.002$ mm) of the analyzed core materials are eroded [27]. The ratio Q_t/Q_0 versus time t is

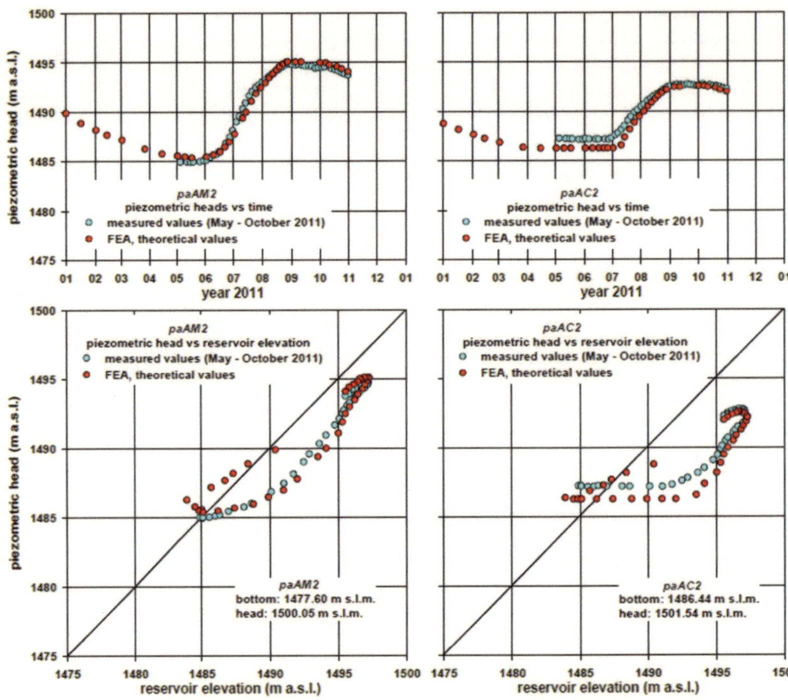

Figure 28. San Valentino dam: measured and theoretical p.h., in 2011 (adapted from Ref. [34]).

Materials	k_0 (m/s)	n_0	φ (°)	c' (kPa)
San Valentino (Base)	1×10^{-6}	0.3	25	0
San Valentino (Transition)	1×10^{-6}	0.3	25	0
Sourva (Base)	1×10^{-6}	0.3	25	0
Sourva (Finer Transition)	2.5×10^{-6}	0.3	25	0
Sourva (Coarser Transition)	5×10^{-6}	0.3	25	0

Table 4. Physical and mechanical parameters characterizing the analyzed materials.

shown in **Figure 29** for the three analyzed case; Q_0 is the initial value of the suspension flow rate: Q_t/Q_0 rapidly increases if the Suorva core material is protected by the coarser transition, due to the intense erosion of the finer fractions of B and the corresponding increase of permeability [27].

After 12 hours, Q_t/Q_0 still increases: both granulometric and hydraulic stabilizations are not yet occurred. If the Suorva core material is protected by the finer transition, after 4 hours Q_t/Q_0 slowly reduces: the erosion of B particles is not still completed; the clogging of the voids of T progressively occurs and the phenomenon seems to achieve a stable state [2]. In San Valentino dam, the washout of the finer particles of B is controlled by the finer protective transition; Q_t

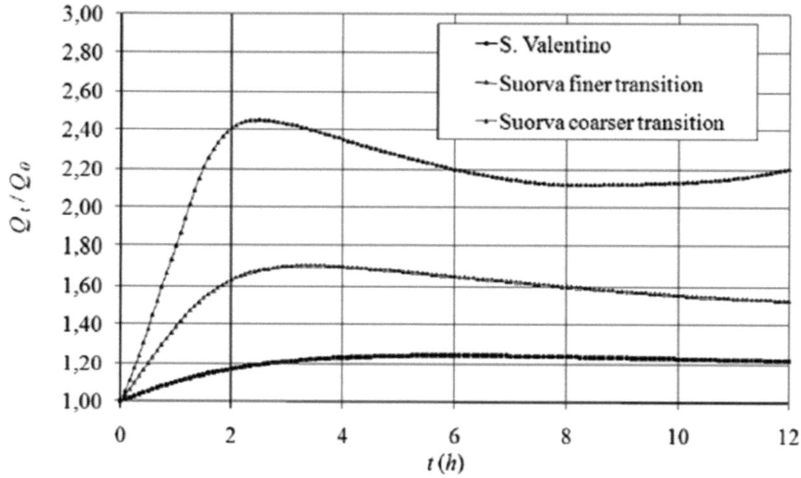

Figure 29. Ratio Q_t/Q_0 vs time *t* (hours) ; Q_0, initial value of the suspension flow rate Q_t (adapted from Ref. [2]).

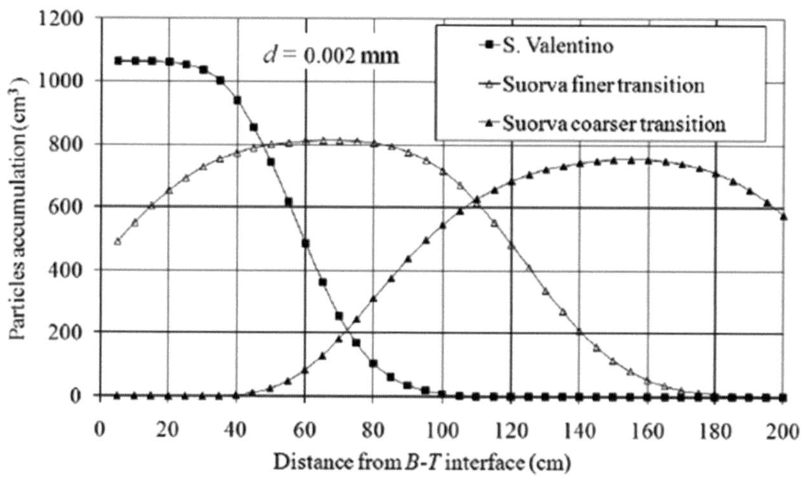

Figure 30. Particles accumulation in the protective transitions (*d* = 0.002 mm) (adapted from Ref. [2]).

slightly increases compared with the Suorva considered cases. Stabilization occurs after 6 hours and Q_t becomes only 1.2 times greater than Q_0.

Very different values of the lengths of the path crossed by the scoured particles, through the analyzed granular transitions (T), are observed (**Figure 30**). Within the T material of the S. Valentino dam, the particles mainly deposit and accumulate just a few cm, after the B-T interface. In the finer T material of Suorva dam, the particles mainly accumulate at about 70 cm after the B-T interface; while, in the coarser transition material of Suorva dam, at about 150 cm after the B-T interface [2].

6. Concluding remarks

An original numerical procedure to simulate the particle migration phenomena in granular media due to seepage flows has been developed. The procedure takes into account voids, constriction sizes, and porosities of the particulate materials (geometric-probabilistic models) as well as the rate of the seeping suspension and piezometric gradients (hydraulic models). The continuous variations (in the space, 1D, and along time, t) of the physical (porosity n) and hydraulic (permeability k) properties of the granular medium, during the coupled seepage flow, deposition and scouring particles processes, are simulated. First, validation of the proposed method has been carried out by simulating selected experimental tests, referred to materials whose diameters range from clay to sand (base materials, B) and sand to gravel (granular transition, T). Then, the proposed numerical procedure has been applied to interpret the empirical Terzaghi's criteria. Finally, the analysis of the granulometric stability of cohesionless moraine materials, constituting the core of earth dams, has been carried out. Through the proposed numerical procedure, the particle migration phenomena occurring at the interface of the contact core–protective transition has been simulated. Results show that the proposed procedure is able to simulate the analyzed deposition–erosion processes; particularly, "*not negligible*" erosion phenomena, involving the finer fractions of these materials, may cause anomalies and malfunctions in dams whose core is constituted by broadly graded cohesionless materials protected by coarse granular transitions.

Author details

Francesco Federico

Address all correspondence to: fdrfnc@gmail.com

Department of Civil Engineering and Information Engineering, University of Rome Tor Vergata, Rome, Italy

References

[1] ICOLD. Internal Erosion of Existing Dams, Levees and Dykes, and Their Foundations. In: Bridle, R. and Fell, R., Eds., Bulletin 164, 2013, Volume 1: Internal Erosion Processes and Engineering Assessment, International Commission on Large Dams, Paris.

[2] Federico F, Montanaro A. Granulometric stability of moraine embankment dam materials: theoretical procedure and back-analysis of cases. In: Proceedings of 6th International Conference on Dam Engineering, C. Pina, E. Portela, J. Gomes (ed.), 15-17 February 2011, Lisbon, Portugal, pp. 423–439

[3] Tran DK, Prime N, Froiio F, Callari C, Vincens E. Numerical modeling of backward front propagation in piping erosion by DEM-LBM coupling. European Journal of Environmental and Civil Engineering 2016, Taylor & Francis ed., DOI: 10.1080/1964189.2016.1248794

[4] Moraci N, Mandaglio MC, Cazzuffi D. I geotessili con funzione di filtro a contatto con terreni granulari: criteri e parametri di progetto. Italian Geotechnical Journal. 2010;**2**:47–71

[5] Litwniszyn. Colmatage-scouring kinetics in the light of stochastic birth-death process. Bull. Acad. Pol. Sci., Ser. Sci. Techn. 1966;14 (9):561

[6] Sakthivadivel R. Theory and mechanism of filtration of non-colloidal fines through a porous medium. Tech. Rep. HEL 15-5, 1966, University of California, Berkeley

[7] Vardoulakis I. Fluidization in artesian flow conditions: Hydromechanically unstable granula media. Geotechnique. 2004, **54**(3):165–177

[8] Saada Z, Canou J, Dormieux L, Dupla JC, Maghous S. Modelling of cement suspension flow in granular porous media. International Journal for Numerical and Analytical Methods in Geomechanics. 2005;**29**:691–711

[9] Zhang XS, Wong H, Leo CJ, Bui TA, Wang JX, Sun WH, Huang ZQ. A thermodynamics-based model on the internal erosion of earth structures. Geotechnical and Geological Engineering. 2013;**31**:479-492. DOI: 10.1007/s10706-012-9600-8

[10] Indraratna B, Vafai F. Analytical model for particle migration within base soil - filter system. Journal of Geotechnical Engineering. 1997;**123**(2):100–109

[11] Musso A, Federico F. Geometrical probabilistic approach to the design of filters. Italian Geotechnical Journal. 1983;**17**(4):173–193

[12] Musso A, Federico F. Pore size distribution in filtration analyses. In: Proceedings of XI ICSMFE, S. Francisco. 1985;**1**:1207–1212

[13] Harr ME. Groundwater and Seepage. New York: Dover Publications Inc., 1962

[14] Kovacs G. Seepage Hydraulics. Elsevier Scientific Publishing Company, Amsterdam; Oxford, New York, 1981

[15] Federico F, Musso A. Some advances in the geometric-probabilistic method for filter design. In: Proceedings of International Conference on "Filters and Filtration Phenomena in Geotechnical Engineering", Karlsruhe, October 1992, pp. 75–82

[16] Silveira A. An analysis of the problem of washing through in protective filters. In: Proceedings of the 6th International Conference on "Soil Mechanics and Foundation Engineering", ICSMFE, Montréal. 1965;**2**:551–555

[17] Silveira A, de Lorena Peixoto T, Nogueira J. On void size distribution of granular materials. In: Proceedings of the 5th Pan-Am Conference on "Soil Mechanics and Foundation Engineering", Buenos Aires, Argentina, 1975, pp. 161–176

[18] Indraratna B, Raut AK, Hhabbaz H. Constriction-based retention criterion for granular filter design. Journal of Geotechnical and Geoenviromental Engineering (ASCE). 2007;**133** (3):266–276

[19] Dallo YAH, Wang Y. Discussion of 'A new theoretical method to evaluate the internal stability of granular soils'. Canadian Geotechnical Journal. 2012;49: 866–868. DOI:10.1139/T2012-036

[20] Moraci N, Mandaglio MC, Ielo D. Reply to the discussion by Dallo and Wang on 'A new theoretical method to evaluate the internal stability of granular soils'. Canadian Geotechnical Journal. 2012;**49**:869–874. DOI: 10.11.39/T2012-047

[21] Wang Y, Dallo YAH. On estimation of the constriction size distribution curve for cohesionless soils. European Journal of Environmental and Civil Engineering. 2014;**18**(6):683–698, DOI: 10.1080/19648189.2014.909335

[22] To HD, Scheuermann A, Williams DJ. A new simple model for the determination of the pore constriction size distribution. In: Proceedings of the 6th International Conference on Scour and Erosion, 31 August 2012, Paris, France, pp. 295–303

[23] Witt K. Filtrationsverhalten und Bemessung von Erdstoff-Filtern. Technical Report 104, D. Inst. F. Bodenmechanik und Felsmechanik, Karlsruhe, 1986

[24] Federico F, Iannucci M. Effects of granulometric stability of cohesionless materials on safety of geotechnical structures. In: Proceedings of International Symposium on Geomechanics from "Micro to Macro" 2014, Cambridge, UK. 2014; **2**, pp. 913–918

[25] Perzlmaier S. Hydraulic criteria for internal erosion in cohesionless soil. Internal Erosion of Dams and their Foundations - Fell & Fry (eds.). London Taylor & Francis Group, ISBN 978-0-415, 2007, pp. 179–190

[26] Biswas S. Study of cohesive soil granular filter interaction incorporating critical hydraulic gradient and clogging. Engineering-Research Master, University Of Wollongong, NSW, Australia, 2005

[27] Federico F, Montanaro A. Numerical analyses of granulometric stability of moraine dam cores. In: Proceedings of 7th European Conference on Numerical Methods in Geotechnical Engineering, Trondheim, Norway, 2010

[28] Atmatzidis DK. A study of sand migration in gravel. In: Proceedings of 12th International Conference on Soil Mechanism and Foundation Engineering., Session 8/3, Rio De Janeiro, Brazil, 1989, pp. 683–686

[29] Skempton AW, Brogan JM. Experiments on piping in sandy gravels. Geotechnique. 1994; **44**(3):449–460

[30] Terzaghi K. Der Grundbruch and Stauwerken and Scine Verhutung. Reprinted in "From Theory to Practice in Soil Mechanics". New York: John Wiley and Sons, Inc.; 1922. pp. 450–476

[31] Terzaghi K. Soil mechanics: a new chapter in engineering science. Journal of the Institution of Civil Engineers. 1939;**12**:106–141

[32] Federico F, Jappelli R, Iannucci M. The Progress of Internal Erosion from Typical Hydromechanical Response of Contacts in Embankment Dams. In: Proceedings of XVI International Technical Conference on Dam Monitoring 2015 – TKZ 2015 - Session 1 "Internal Erosion in Earth Structures – Poland, 29 September – 2 October

[33] Nilsson A. The susceptibility of internal erosion in the Suorva dam. Internal Erosion of Dams and their Foundations, Editors R Fell and J-J Fry, Taylor and Francis; 2007. pp. 167–172.

[34] Pinamonti P, Scienza M, Catalano A, Del Gizzi F, Jappelli R, Federico F, Montanaro A. Safety of the San Valentino Earth Dam after 60 Years of Operation. In: Proceedings of 8th ICOLD European Club Symposium 2010 - DAM SAFETY, Sustainability in a Changing Environment, 22–23 September 2010, Innsbruck, Austria

[35] Federico F, Jappelli R, Montanaro A, Scienza M. Self-protection of a quasi-homogeneous embankment dam revealed by advanced analyses and monitoring. In: Proceedings of 8th European Conference on Numerical Methods in Geotechnical Engineering, June 18-20, 2014, Delft, The Netherlands

Chapter 7

The Grading Entropy-Based Criteria for Structural Stability of Granular Materials and Filters

János Lőrincz, Emöke Maria Imre and Vijay Pal Singh

Additional information is available at the end of the chapter

http://dx.doi.org/10.5772/intechopen.69167

Abstract

Some rules for particle migration, filtering, and segregation were elaborated on the basis of some simple laboratory tests and data of well-designed, artificial mixtures of natural sand grains. Use was made of the knowledge available in the field and two pairs of grading entropy parameters. These parameters incorporate all information of the grading curve and are pseudo-metrics in the "space of the possible grading curves."

Keywords: grading entropy, internal erosion, suffosion, filtering, segregation, piping

1. Introduction

The internal stability of compacted earth dam materials, granular filters, and soils on natural slopes is essential. The internal erosion involves loss of particles under seepage flow; the matrix of coarse soil particles may or may not be unstable [1–3]. The term "suffosion" is Russian in origin and is used to describe the process of removal and transport of small soil particles through pores [4, 5].

It is desirable that adjacent materials in earth dams or rockfill dams should act as filters for each other and the material should not segregate [6–14]. Broadly graded materials may segregate during the construction process where the particles are able to flow freely, such as tipping and spilling. The likelihood of backward erosion is greater for segregated soil than for non-segregated soil [7, 8].

The inherent stability or proneness to segregation is usually specified in terms of particular diameters D_x (or d_x), which represent the particle diameter for which x% of grains (by weight)

are smaller. The susceptibility to suffosion is assessed by the graphical approach [5], where a grain size distribution is compared with empirical upper- and lower-bound thresholds; the method is not valid for gap-graded grading. The filter rules—the compacted earth dam or core material should obey when associated with each other and with the dam base—are formulated in terms of pairs of grading curve points [9–15], which is "too simple" in case of broadly graded soils.

This chapter summarizes three grading entropy-based rules, on the basis of the original work of Lőrincz and some applications [15–22]. The suggested rules differ from most existing rules in that the whole grading curve is used instead of some limited number of grading curve points, without any constraint on the shape of the grading curve. They were elaborated on the basis of the knowledge available, the measured data available in the literature, and data measured for well-designed sand mixtures by Lőrincz. The rules were verified by the examined cases [19–22], an example included.

2. Grading curve characterization

The grading curve is a statistical distribution of logarithm of the diameter with respect to the dry weight. It is a discrete distribution curve with a non-uniform cell system in arithmetic scale. To characterize it, first of all, the statistical cell system—the so-called abstract fraction system—is defined and the space of the grading curve is introduced.

Then the two grading entropy parameter pairs are introduced. The first pair is related to the expected (log diameter) value of the grain size distribution, in non-normalized and normalized forms. The normalized version has a shift symmetry on the log d axis.

The second pair is the entropy arisen from the mixing of the fractions, in non-normalized and normalized forms. Its maximum for a fixed value of the first coordinate is related to a single grading curve with finite fractal distribution. The grading curves can be represented in terms of the two parameters in the entropy diagram.

2.1. The fractions

The fraction system is defined on the pattern of the classical sieve hole diameters (where measurements are made), by successive multiplication with a factor of 2, as follows. The diameter range for fraction j ($j = 1, 2\ldots$, see **Table 1**):

$$2^j d_0 \geq d > 2^{j-1} d_0, \tag{1}$$

or the upper diameter range for fraction j:

$$d_j = 2d_{j-1} \tag{2}$$

Using $\log_2$ form results in an integer increment by each multiplication and fraction as follows:

http://dx.doi.org/10.5772/intechopen.69167

$$\log_2 d_j = 1 + \log_2 d_{j-1} \tag{3}$$

Fraction number j	1	...	23	24	...
d [–]	d_0 to $2\ d_0$	...	$2^{22}\ d_0$ to $2^{23}\ d_0$	$2^{23}\ d_0$ to $2^{24}\ d_0$	...
d [mm]	2^{-22} to 2^{-21}	...	1–2	2–4	...
S_{0j} [–]	1	...	23	24	

Table 1. Definitions of fractions, based on the smallest particles likely to occur in nature.

The variable d_0 is the arbitrary smallest diameter, and assumingly a 2-power, $d_0 = 2^{-k}$. Its possible value is equal to the height of the SiO_4 tetrahedron ($d_0 = 2^{-22}$ mm).

$$\log_2 d_1 = 1 + -22 \tag{4}$$

The fraction serial number variable can be expressed by the diameter:

$$j = k + \log_2 d_j \tag{5}$$

The integer $j/j - 1$ is a so-called abstract upper/lower diameter limit ($\log_2 d_j$ shifted by k).

2.2. The grading curve space

By the measurements of the fractions during sieving, the relative frequencies of the fractions x_j ($j = 1, 2, 3\ldots N$) can be determined. These fulfill the following equation of each grading curve:

$$\sum_{j=1}^{\infty} x_j = 1, x_j \geq 0 \tag{6}$$

which can be rewritten as follows:

$$\sum_{i=1}^{N} x_i = 1,\ x_i \geq 0,\ N \geq 1. \tag{7}$$

where i is a rescaled fraction serial number being equal to 1 at the finest non-zero fraction with original serial number j_{min} (see **Table 1**), the integer variable N is the number of the fractions between the finest j_{min} and coarsest j_{max} non-zero fractions:

$$N = j_{max} - j_{min} + 1. \tag{8}$$

The space of the grading curves with N fractions can be identified with the $N - 1$-dimensional, closed simplex (which is the $N - 1$-dimensional analogy of the triangle or tetrahedron, the two- and three-dimensional instances, **Figure 1**) as follows. Each grading curve is related to a simplex point, the relative frequencies x_i can be identified with the barycentric coordinates of the points of the $N - 1$-dimensional closed simplex.

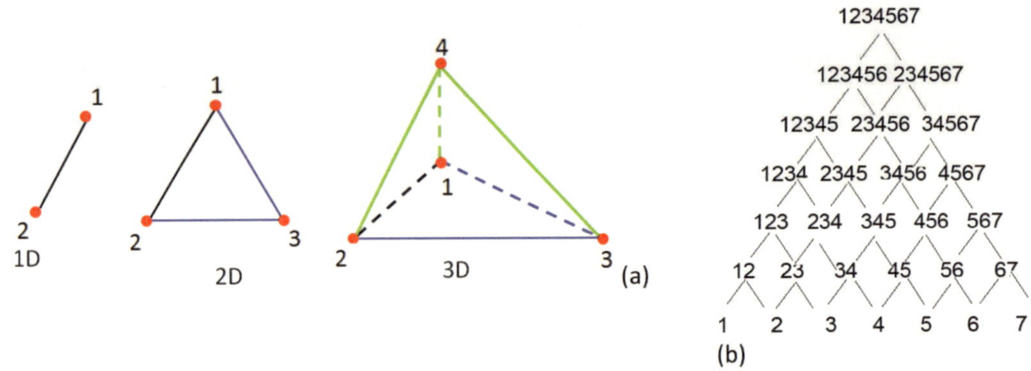

Figure 1. (a) Standard simplex images with dimension less than 4. (b) The lattice of the continuous sub-simplexes of the six-dimensional simplex.

2.3. The grading entropy coordinates of the grading curve

The grading entropy concept is an application of the statistical entropy to the grading curve [15, 23], by introducing a uniform cell system for the derivation besides the fractions. It condenses the information of the whole grading curve into two pairs of parameters. The grading entropy S is the sum of two "means" [15]:

$$S = S_0 + \Delta S \tag{9}$$

which are called as base entropy S_0 and entropy increment ΔS. The base entropy S_0 is a weighted mean or expected value of the fraction serial number:

$$S_0 = \sum_{i=i_{\min}}^{i_{\max}} x_i S_{0i} \tag{10}$$

which depends linearly on the $\log_2$ diameter d (see Eq. (5)), S_{0i} is the grading entropy of the ith fraction, being identical to the fraction serial number (see **Table 1**):

$$S_{0i} = i \tag{11}$$

Any decrease in the base entropy S_0 can be explained by the decrease of the mean grain diameter, for example, due to breakage. Any increase in the base entropy S_0 can be explained by the increase of the mean grain diameter, for example, due to suffosion or segregation.

The relative base entropy A is defined as follows:

$$A = \frac{S_0 - S_{0\min}}{S_{0\max} - S_{0\min}} \tag{12}$$

where $S_{0\max}$ and $S_{0\min}$ are the entropies of the largest and the smallest fractions in the mixture, respectively.

The relative base entropy A varies between 0 and 1, its value is equal to 0.5 if all relative frequencies of the fractions x_j ($j = 1, 2, 3,\ldots, N$) are equal. It measures the distance of the mean $\log_2$ diameter and the smallest $\log_2$ diameter. Geometrically, the grading curves with the same A have the same sub-graph area. Since A is linear, the A = constant condition in addition means parallel hyperplanes in the $N - 1$-dimensional space generated by the simplex (**Figure 2(a)** and **(b)**).

The entropy increment ΔS is the logarithm of the weighted generalized geometric mean of the relative frequencies of the fractions x_j ($j = 1, 2, 3,\ldots, N$):

$$\Delta S = -\frac{1}{\ln 2}\sum_{x_i \neq 0} x_i \ln x_i. \tag{13}$$

The entropy increment ΔS measures how much the soil behavior is really influenced by all of its N fractions. For those grading curves, in which all N fractions are well represented, the entropy increment is typically close to $\ln N/\ln 2$. The normalized entropy increment B is defined as follows:

$$B = \frac{\Delta S}{\ln N} \tag{14}$$

Being a strictly concave function, the normalized entropy increment B has a unique maximum for each A = constant value, which is the following grading curve with finite fractal distribution (see the definition in [24]):

$$x_1 = \frac{1}{\sum_{j=1}^{N} a^{j-1}} = \frac{1-a}{1-a^N}, \tag{15}$$

$$x_j = x_1\, a^{j-1} \tag{16}$$

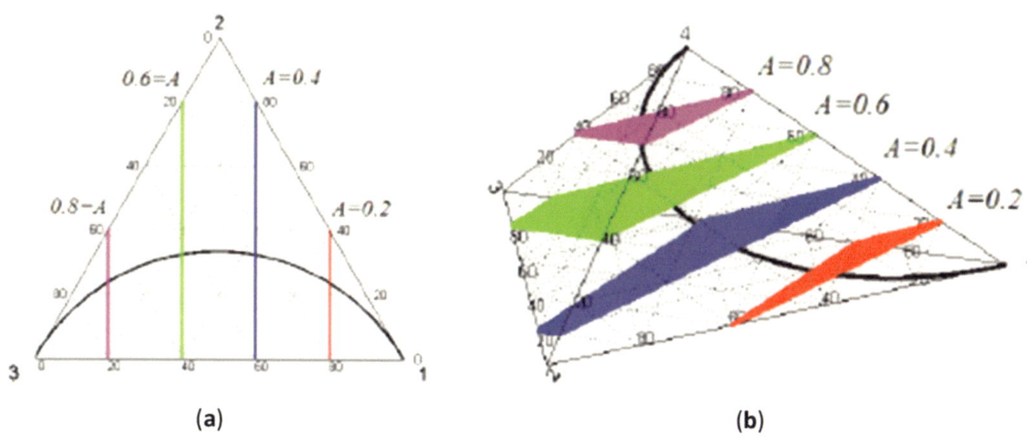

Figure 2. The constant A sections of the simplex and the optimal line (a) for a two-dimensional simplex and (b) for a three-dimensional simplex [16].

where parameter a is the root of the following equation:

$$y = \sum_{j=1}^{N} a^{j-1}[j - 1 - A(N-1)] = 0. \tag{17}$$

As A varies between 0 and 1 (the extreme values represent the extreme fractions 1 and N), the positive root a varies between 0 and ∞ in the function of N. The optimal grading curve has finite fractal distribution with fractal dimension n given by:

$$a = 2^{(3-n)} \tag{18}$$

The optimal grading curve is concave if $A < 0.5$, linear if $A = 0.5$, convex if $A > 0.5$ (see **Figure 3**). Having no inflexion points, the optimal grading curve has the shortest curve length out of the possible grading curves with a specified A. The optimal points of the simplex constitute a continuous line called optimal line which can be seen in **Figure 2**.

In the linear case, it has a unique maximum, being equal to 1/ln 2, in the center of the simplex where each relative frequencies x_i are the same, in this case the "disorder" is maximal in the system. The disorder originated from mixing of the fractions can be measured by the entropy increment ΔS, which is the entropy of the fractions neglecting the fact that the width of the statistical cells in arithmetic scale is different

2.4. The entropy diagram

Four kinds of maps can be defined between the $N - 1$-dimensional simplex with fixed N, with fixed smallest fraction serial number $i_{\min}$ and the two-dimensional space of the entropy coordinates: entropy map with coordinates $[S_o, \Delta S]$, the normalized entropy map with coordinates $[A, B]$ and the two partly normalized entropy maps with a mixture of normalized and non-normalized coordinates, that is, $[S_o, B]$ or $[A, \Delta S]$. The optimal line of the simplex—between vertices 1 and N—

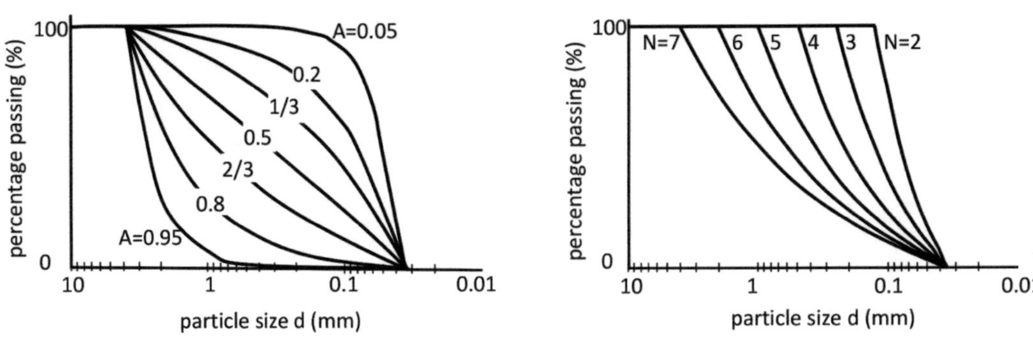

Figure 3. Optimal grading curves with finite fractal distribution, (a) $N = 7$. A varies. (b) $A = 2/3$, N varies.

These maps are continuous on the closed simplex for fixed N. The image of the simplex—the entropy diagram—has a maximum and a minimum value for every possible value of A or S_o. The optimal line of the simplex maps into the maximum line, the map along the optimal line—maximum B line is one to one (see **Figure 4(c)** and **(d)**). The simplex edges and vertices map into the minimum line. The minimum and maximum diagram lines—illustrated in **Figures 4** and **5**—differ in scaling.

In terms of the original entropy coordinates, the map is continuous if N is changing, and in terms of the normalized entropy coordinates, the map is not continuous if N is changing. The normalization with respect to a coordinate results in the range being fixed in that direction. The maximum B point is equal to $l/\ln 2 = 1.44$ for any N, the maximum of the entropy increment ΔS is equal to $\ln N/\ln 2$.

The images of the optimal lines—the maximum B lines—will nearly coincide for any number of the fractions N (**Figure 5(b)**) in the normalized diagram but will separate in terms of the non-normalized coordinates [S_o, ΔS], reflecting the structure of the continuous sub-simplexes of a simplex. This is illustrated in **Figures 1(b)** and **5(a)** for soils with up to seven fractions.

The inverse image of a regular normalized entropy diagram point [A,B] in the simplex is situated on the A = constant, $N - 2$-dimensional, affine hyperplane, with shape of an $N - 3$-dimensional

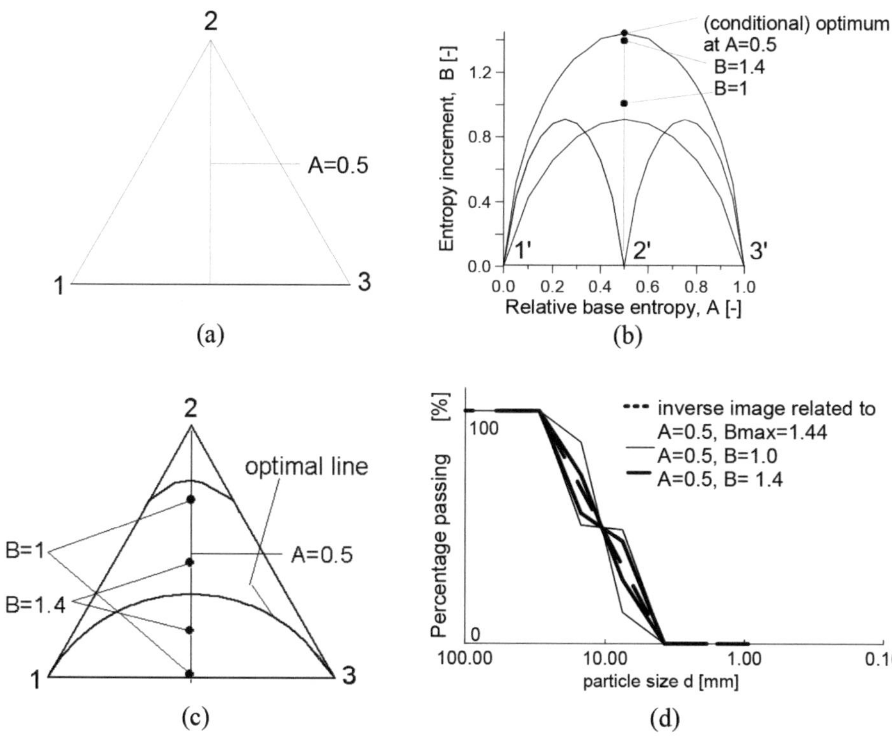

Figure 4. The normalized entropy map and the inverse image for $N = 3$. (a) The two-dimensional simplex and its $A = 0.5$ hyperplane section. (b) The image of the simplex in the entropy diagram. (c) and (d) The inverse image of points $A = 0.5$, $B = 1$, 1.4 and $B_{max} = 1.44$ in the simplex and in the space of the grading curves, respectively.

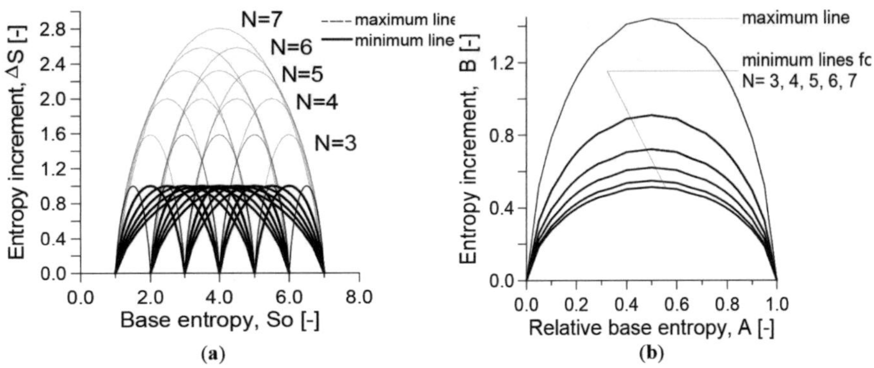

Figure 5. Diagrams of simplexes with various N, the image of the optimal lines ("maximum lines") and the image of the edges $1 - N$, $N = 2–7$ (i.e., "minimum lines"). (a) The non-normalized diagram. (b) The normalized diagram.

"sphere," centered to the optimal point, its "radius" depends on $B_{max} - B$. The related grading curves have the same sub-graph area and the deviation from the optimal grading curve depends on $B_{max} - B$.

3. The construction of the grading entropy-based rules

3.1. The methods

The particle migration (or internal stability) rule, the filter rule, and the segregation rule were constructed as follows. For each rule, simple soil testing programs were designed and executed by using artificial mixtures of natural sand grains [15]. Two variables were carefully constructed using the grading entropy concept [15] for each rule separately.

The entropy variables were used such that the experimental data were plotted on diagrams, differentiating points which exhibited different physical behavior so that domains of particular behaviors could be defined. In addition, some other information and existing data (e.g. the data base related to Ref. [7]) were used.

The additional information was as follows: one piece of information used was that there could be no more than two empty particle size fractions between the filter and the base soil, before the base soil cannot be retained by the filter [21]. This can be derived using Pure Geometry Theorems [21] and also by using the Terzaghi filter criterion (i.e., the finer is to be protected) as follows:

$$1 \le \frac{D}{d} \le 4 \tag{19}$$

where D and d are the diameter of the filter and the base soil. Another piece of information was that the data of suffosion tests could be used for the filter rules, and vice versa, based on the following consideration which comes from the self-filtering theory of Kézdi [25, 26]. If the gap-graded grading curve (see eg. **Figure 6**) is cut into two parts, with the coarser part

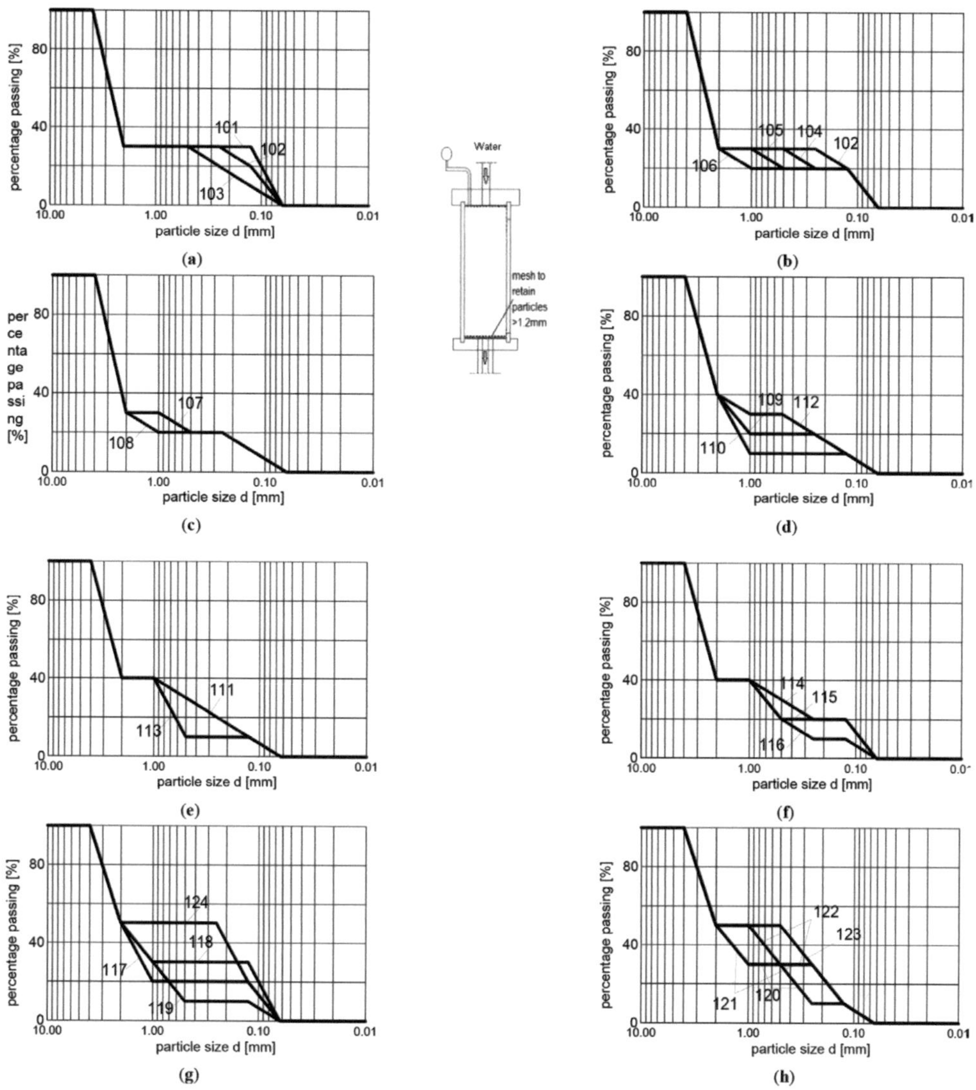

Figure 6. (a–h) Some grading curves of samples used by Lőrincz [15, 21] for suffosion tests. The inset shows the permeameter test arrangement used in the tests.

considered as the filter and the finer part as the base soil, and the base soil part being filtered, then there will be no suffosion within the soil.

3.2. Particle migration rule

For the particle migration (internal stability-suffosion) rule, simple vertical flow tests were designed and executed using artificial mixtures of natural sand grains. The dimensions of the permeameter were 20 cm in height and 10 cm in diameter. It was closed at the bottom by a sieve which was permeable of grains smaller than 1.2 mm but which retained grains larger than 1.2 mm.

The downward hydraulic gradient [*i*] was between four and five. The two parts of the permeameter were separated after the test, and the grading curves were determined. The grain movements were detected.

The results of the vertical water flow (suffosion) test were represented in the partly normalized entropy diagram, in terms of the relative base entropy and the entropy increment coordinates A and ΔS, as shown in **Figure 7(a)** and the rule was separately set up for each specified N value. The part of the diagram where the gap-graded grading curves resulted in self-suffosion was indicated by letter *b*. The simplified normalized diagram version is shown in **Figure 7(b)**.

If $A < 2/3$ (zone I), the soil was internally unstable; if $A = 2/3$ and $A < 2/3$ (zones II, III), the soil was internally stable. In zone II, there were no particle movements: the larger particles retained the smaller particles and the smaller particles supported the larger ones. In zone III, the fines may have migrated in the presence of seepage flow ("suffosion").

The internal stability zone III was separated by the 2/3 vertical line. The division curve between zones II and III connects the maximum entropy points of the mixtures with fraction number less than N, the shape of the curve between II and III is approximate in **Figure 7(b)**.

The rule can be interpreted such that, in zone I (where $A < 2/3$), no structure of the larger grains is present, the coarse particles "float" in the matrix of the fines and become destabilized when the fines are removed by piping. In the zone where $A = 2/3$ and $A > 2/3$ the coarse particles form a skeleton and total erosion cannot occur. In zone III, the structure of larger particles is inherently stable, the smaller particles may move by suffosion.

3.3. Filter rule

The filter rule was developed using three series of tests: the filter tests of Sherard [7], the filter tests of Lőrincz [15], and the suffosion tests of Lőrincz [15]. The grading of the soils tested by

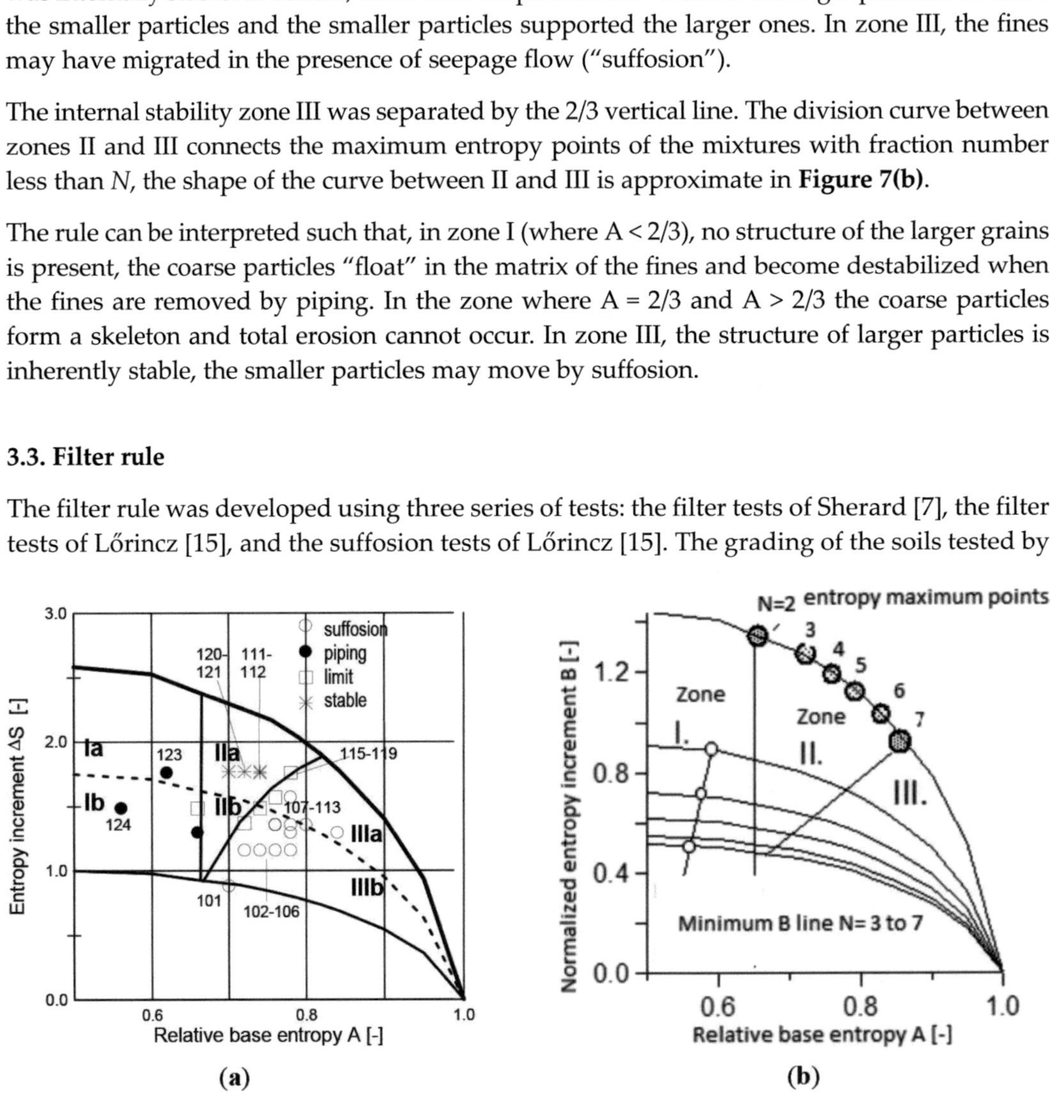

Figure 7. (a) Particle migration zones in half of the partly normalized entropy diagram for mixtures with $N = 6$ fractions, the three digit numbers are related to the grading curves shown in **Figure 6**. (b) Particle migration zones in the simplified, normalized entropy diagram [21].

http://dx.doi.org/10.5772/intechopen.69167

Sherard is shown in **Figure 8** and the grading of the soils tested by Lőrincz is shown in **Figures 6** and **9**. In the filtering test, the filter and base soils are placed into the permeameter (20 cm in height and 10 cm in diameter) in series separated by a sieve. The downward hydraulic gradient [*i*] is between four and five. The suffosion tests of Lőrincz [15] were reanalyzed as follows. The gap-graded grading curves shown in **Figure 6** were cut into two parts at the gap and one part was considered as the filter and the other part as the base soil. If suffosion occurred, then the filter was unsafe.

Two pseudo-metrics were constructed from the grading entropy parameters. The logarithm of the difference between base entropies of the filter and base soils, $\log(S_{0f} - S_{0b})$ described the distance between the mean $\log_2$ diameters of the filter and base soils. The sum of filter and base soil entropy increments $\Delta S_f + \Delta S_b$ expressed the sum of the two "effective" fraction number or N values (i.e., the total number of important fractions in the two grading curves).

Plotting the test results in terms of the foregoing variables, the safe and unsafe areas were separated by a straight line: a layer acts as a filter for an adjacent layer (for the base soil) on the condition that:

$$\Delta S_f + \Delta S_b \geq 4\ln(S_{0f} - S_{0b}) - 4.39 \tag{20}$$

where index f and b denote the filtering layer and the material being filtered (base soil), respectively. The domains defined by the Eq. (20) are shown in **Figure 10**. The point where

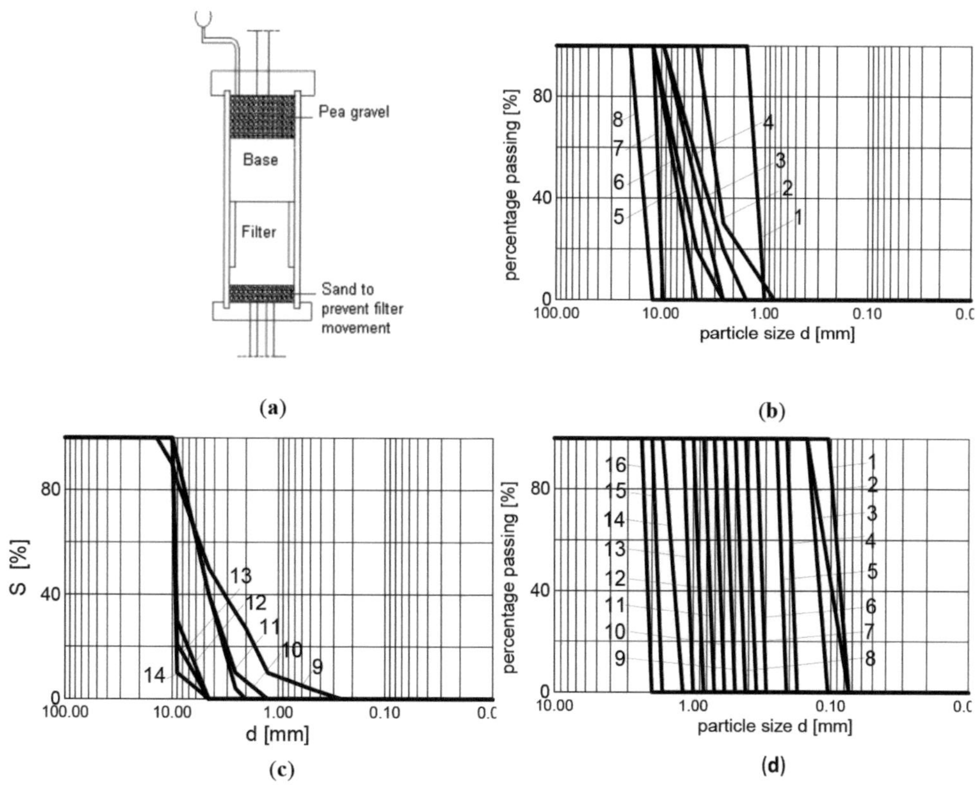

Figure 8. (a) Filter test apparatus in [7, 21]; (b) and (c) Sherard-filter soils [7, 21]; (d) Sherard-base soils [7, 21].

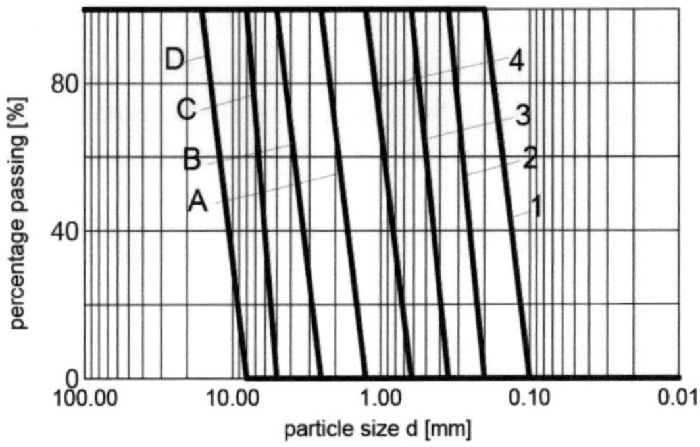

Figure 9. Some grading curves of soils used in the filter tests of Lőrincz [15, 21]. Note that the filters are identified by characters, and the base soils by numbers.

$\Delta S_f + \Delta S_b = 0$ was defined on the basis of the assumption that in the limit state two empty size fractions do exist between the filtering fraction and the filtered fraction.

3.4. Segregation rule

For the segregation rule, the simple e_{max} test was used in a modified form. In the e_{max} tests, the soil is poured into a funnel and flow is allowed from the funnel into a cylinder (10 cm in height and 10 cm in diameter). The segregation test was made with about double the quantity that expectedly filled the 10 cm diameter, 10 cm high cylinder. The funnel is just rising above the soil surface.

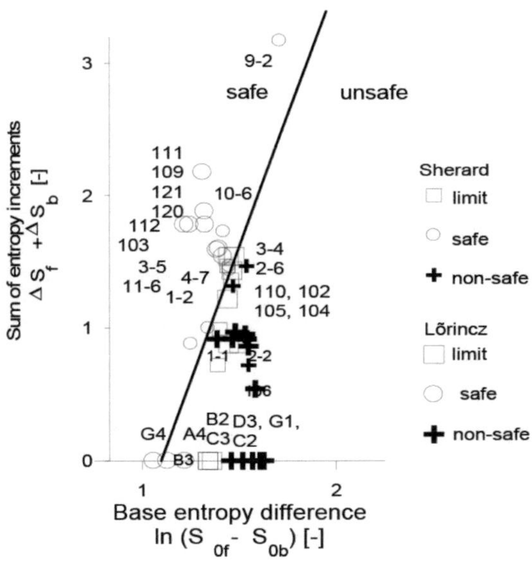

Figure 10. The filter rule with the safe and unsafe areas. The soils are shown in **Figures 8** and **9** [15, 21].

The artificial mixtures of natural sand grains used for the segregation rule were partly continuous mixtures (A, B), partly gap-graded mixtures (C, D, E), as shown in Ref. [21]. The results are shown in **Figure 11**. The difference in the initial and the poured base entropy S_0, entropy increment ΔS, and grading entropy S was represented in the function of the relative base entropy A. According to the results, some segregation was always measured, but it was with minor significance if the relative base entropy of the soil was within the following limits:

$$0.4 \leq A \leq 0.7 \tag{21}$$

The results showed that the base entropy difference S_0 was negative and the entropy increment difference ΔS was positive above the lower limit $A = 0.4$. Physically, the proportion of the large grains and the mean $\log_2$ diameter is larger if the base entropy difference S_0 is larger and vice versa. The value of ΔS is larger if the mixture is more uniform in terms of fraction relative frequencies.

3.5. Applications

3.5.1. Non-segregating mixtures

Minimal segregation occurs, and relatively uniformly textured body of soil is achieved for laboratory testing of granular materials or for the earth works, if a non-segregating mixture with $0.4 < A < 0.7$ is used.

To construct continuous, non-segregating mixtures, some optimal limit curves can be determined. The optimal mixtures computed by a simple algorithm fulfilling Eqs. (15)–(17) for fixed $N = 5$ (where B is a maximum for a given A are for five fractions) are shown in **Figure 12** and **Table 2**. Similar limit curves can be reproduced for any fraction number.

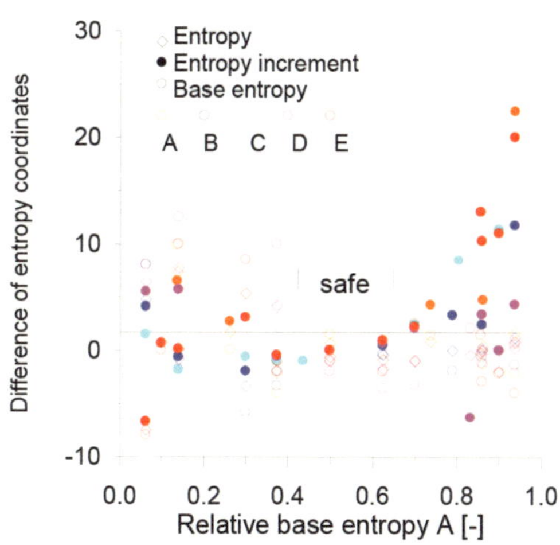

Figure 11. Results of the segregation test, mixtures A–E are shown in [21].

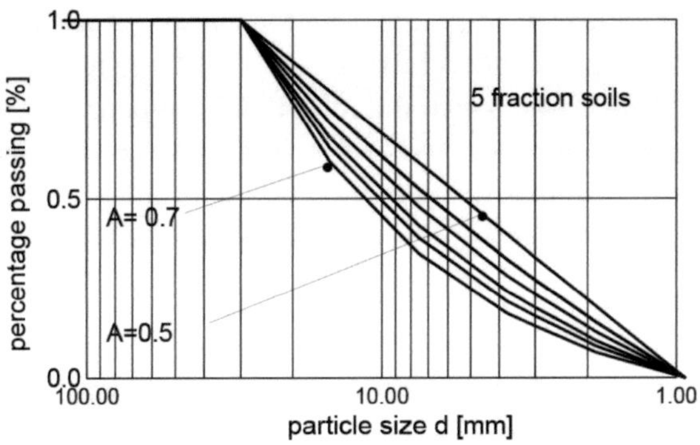

Figure 12. Limit curves for non-segregating optimal 5-fraction mixtures [21].

A [–]	a [–]	x_1 [–]	$\Delta S/\ln N$ [–]
0.50	1.00	0.20	1.44
0.56	1.13	0.15	1.43
0.60	1.23	0.13	1.41
2/3	1.42	0.09	1.34
0.70	1.54	0.07	1.29

Table 2. Some non-segregating optimal 5-fraction mixtures.

However, it can be noted that soils with gap-graded grading curves can be non-segregating also. For example, in case of a two-fraction soil with gap-graded grading curve, the segregation is minimal if the quantity of the larger fraction varies between 0.4 and 0.7.

3.5.2. Testing the filter rules

The grading entropy-based filter law was compared with the existing filtering rules available in the literature. Summaries of well-known filter rules [7, 9–15] for uniformly graded filters and broadly graded filters are presented in Appendix A. These different filtering rules were tested by generating soils with the special-shaped grading curves [21] shown in **Figure 13** and parameterized in **Table 3**.

The 13 combinations listed in **Table 3** were represented for the different filtering rules of the literature, some results are shown in **Figure 14**. If the rule from the literature predicted a successful filtering (i.e., safe behavior), it was plotted with an open circle; where it predicted a failure to filter (i.e., unsafe behavior), it was plotted with a full circle.

The results indicated that (i) the Terzaghi's filter rule is too conservative, (ii) the Bertram rule is conservative for mixed filters and not acceptable for uniform soils, (iii) the rule of United States Bureau of Reclamation (USBR) for uniform filters is acceptable, and (iv) for mixed filters is not acceptable.

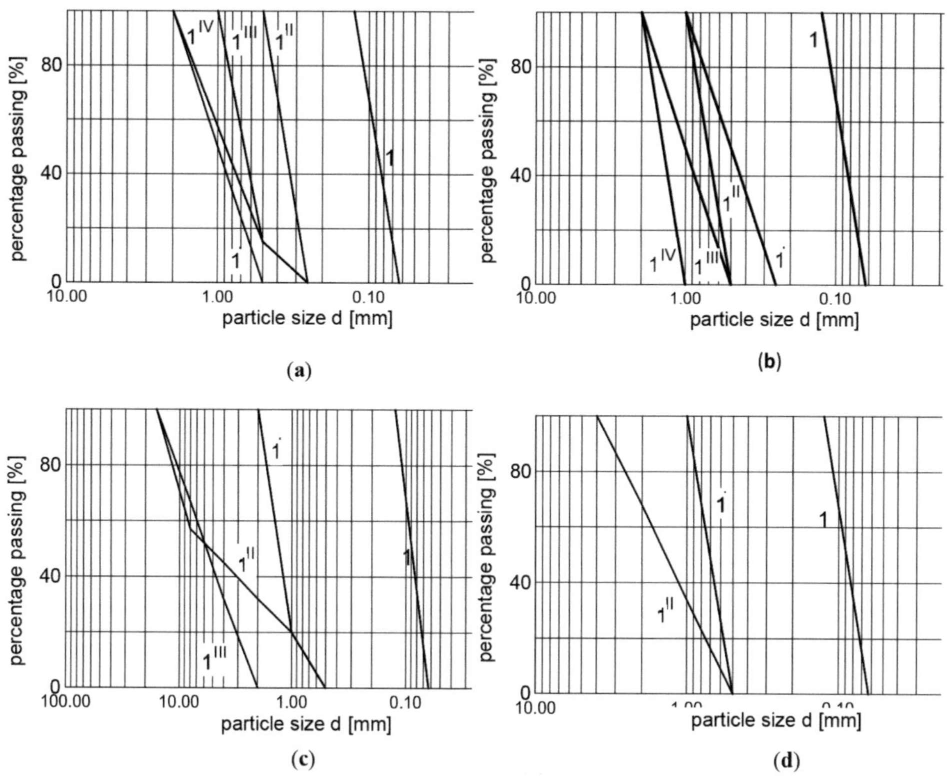

Figure 13. The theoretical grading curves used in the testing of existing filtering laws. (a) Mixtures for the Terzaghi's criterion-T. (b) US Bureau simple filters-U. (c) US Bureau mixed filters-UM. (d) Mixtures for Bertram's criterion-B.

		D_{50}/d_{50}	D_{15}/d_{85}	D_{10}/d_{60}	D_{15}/d_{15}	S_{0b}	S_{0f}	ΔS_b	ΔS_f
1	B1-1^{I}	7	4.58	6.5	7.86	13	16	0	0
2	B1-1^{II}	14	5.67	22.5	9.71	13	17	0	1.585
3	T1-1^{I}	10.00	4.17	4.44	7.14	13	16.35	0	1.44
4	T1-1^{II}	4	2.42	3.1	4.14	13	15	0	0
5	T1-1^{III}	6.89	4.17	3.1	7.14	13	15.85	0	0.61
6	T1-1^{Iv}	11.11	5.17	6.3	8.86	13	16.5	0	1
7	UM1-1^{I}	13.9	7.3	0.8	12	13	16.8	0	0.722
8	UM1-1^{II}	58	7.3	0.8	12	13	18.51	0	2.07
9	UM1-1^{III}	58	25.5	2.6	40	13	19.	0	1.585
10	U1-1^{I}	5.55	3.02	3.13	4.57	13	15.5	0	1
11	U1-1^{II}	7.78	5.19	5.42	7.86	13	16	0	0
12	U1-1^{III}	11.1	5.85	6.25	8.86	13	16.5	0	1
13	U1-1^{Iv}	15.5	10.19	10.63	15.43	13	17	0	0

Table 3. The data of the theoretical grading curves used in evaluating the filtration rules, generated for this purpose [15] to four existing filter rules, as shown in **Figure 13**.

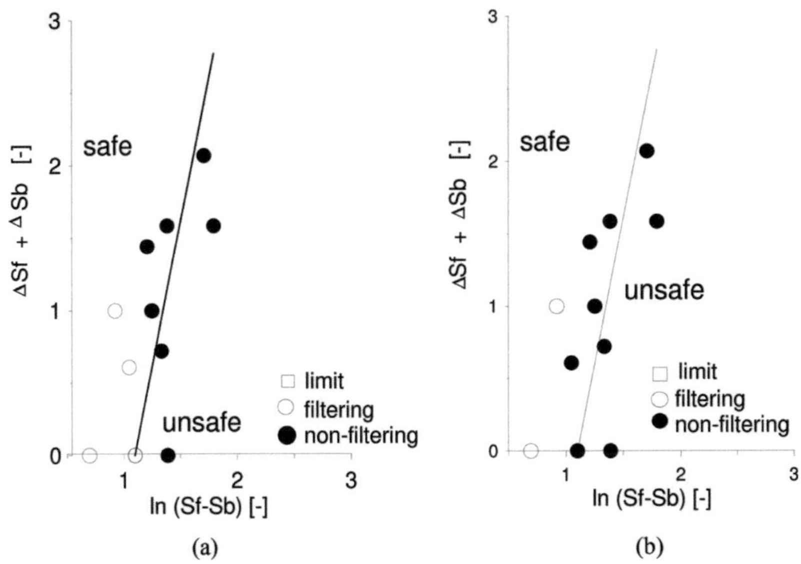

Figure 14. Testing some filtering rules, using the theoretical grading curves of **Figure 13** and **Table 3**. (a) The filter rule of USBR for uniform filters [12] is acceptable. (b) The filter rule of Terzaghi [11] is conservative.

4. Case study

Several applications of the entropy-based rules, by examining the reason of piping, softening, dispersive soil behavior, and the goodness of a leachate collection system, were previously presented [19–22]. Here, a dam failure case study is summarized.

The 71 m high, Gouhou rockfill dam was founded on a sandy gravel base layer (**Figure 15**). The dam body consisted of the following parts: the upstream face was a thin layer of material with a design particle diameter of 100 mm, zone I was a transition zone with the maximum diameter of 400 mm, zones II and III were the main rockfill with maximum diameters d of 600 and 800 mm.

The dam failed [22, 27, 28], killing 288 people, immediately after the first rising water level, and infiltrating the water into the dam body causing internal erosion, piping, and washout of material (see 1–6, **Figure 15**).

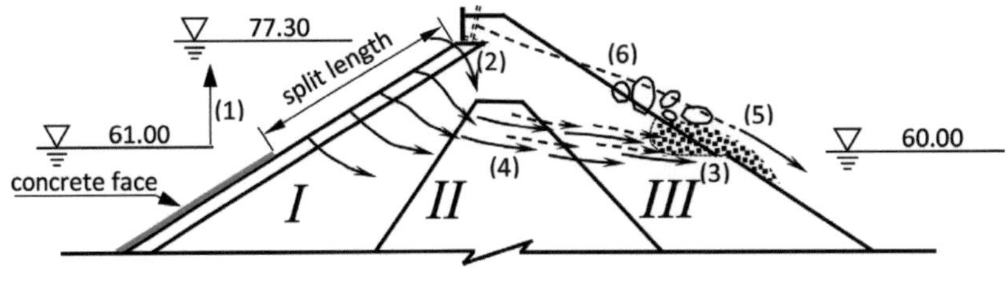

Figure 15. The Gouhou dam failure. Cross section and failure mechanism.

The relative base entropies of the soils in zones I, II, and III were 0.42, 0.55, 0.58, respectively, all less than 2/3 and non-segregating. This result explains why the rockfill material was incapable of forming a stable skeleton of coarse fragments. It follows that the grading entropy-based soil behavior rules would have been capable of predicting piping failure in the Gouhou dam.

5. Discussion and conclusion

5.1. Some comments on the entropy parameters

The grading entropy parameters are some kind of integrals of the whole grading curve. The same shaped grading curve has the same A, ΔS, or B value independent of the value of the minimum grain size. Therefore, these are well-defined parameters and have some physical contents, as follows.

The relative base entropy parameter A has a potential to be a grain structure stability measure possibly based on the simple physical fact that if enough large grains are present in a mixture, then these will form a skeleton.

The entropy increment ΔS measures how much the soil behavior is really influenced by all of its N fractions. For those grading curves, in which all N fractions are well represented, the entropy increment is typically close to $\ln N/\ln 2$.

The entropy parameters are pseudo-metrics. The difference between base entropies of the filter and base soils, $S_{0f} - S_{0b}$ describes the distance between the mean diameters. The sum of filter and base soil entropy increments $\Delta S_f + \Delta S_b$ expresses the sum of the two "effective" N values (i.e., the total number of important fractions in the two grading curves).

5.2. Some comments on the rules

The overall soil stability—according to the experimental results—is described by the criterion that $A > 2/3$. In soils which meet this criterion, the matrix of coarser soil particles is stable and able to form a resistant skeleton, even though suffosion may occur.

Some questions arise, for example, in regard to the stability of a single fraction which does not lie in a unique position on the entropy diagram. Since the change due to degradation is the appearance of smalls, which causes an increase in the A value [29], the one fraction case is likely on the safe side.

Another question is related to the probability that an arbitrary N-fraction soil is stable. This can be characterized by the relative size of the grading curve space separated with the $A = 2/3$ hyperplane on condition that the probability is the same in the whole simplex. This number is decreasing with the fraction number (e.g., for $N = 2$, the 1/3 part of the grading curve space is safe, for $N = 3$, the 2/9 part of the possible grading curves are safe, for $N = 18$, less than the 0.01 part of the possible grading curves are safe).

Significant segregation is unlikely to occur, if the relative base entropy A is between the limits of 0.4 and 0.7. It is important to note that the same parameter—the relative base entropy A—is responsible for overall soil stability. Soils which meet both criteria may constitute very small part of the grading curve space and may need careful design in case of broadly graded soils.

The filtration problems are safely solved in the literature for uniform filters and bases (i.e., soils to be protected by the filters). The suggested filter rule can be used to design broadly graded filters (e.g., for clay cores or for leachate collection systems). However, the rule was estimated on the basis of one data point only at the range of very large N and ΔS values.

The testing of the existing rules known from the literature was possible on the basis of the suggested filter rule and using some theoretical grading curves. According to the results, the Terzaghi filter rule is too conservative. The filtering rule of USBR for uniform filters is acceptable. The mixed filter rule of the USBR is not conservative and is not acceptable. The Bertram rule is be conservative for mixed filters and not acceptable for uniform soil.

5.3. The importance, use, and implementation of the rules

Applications of the derived entropy-based rules were presented by examining the reason of a dam piping failure, dike piping, dispersive soils, leachate collection system case studies [19–22], a dam example is presented here only. On the basis of the case study, it is apparent that the grading entropy-based soil behavior rules would have been capable of predicting piping failure in the Gouhou dam.

The grading entropy-based criteria can easily be implemented into any laboratory test evaluation software. A basic requirement for the use is that the grading curve information is reliable. The simple soil tests presented here were made on coarse material and the rules apply for soils where the solid fraction is composed of non-clay minerals.

For clay minerals, the same criteria may be valid if the grading curve information is reliable and the appropriate degree of particle agglomeration is reflected in the measurements [30–33]. The first results indicate that the same criteria may be valid for silty soils if the grading curve information is reliable (see e.g., the dispersive soil case studies [20]).

Appendix A

Rules for uniform filters

U.S. Bureau of Reclamation [12]:

$$\frac{D_{50}}{d_{50}} = 5 - 10 \tag{22}$$

Sichard [9]:

$$\frac{D_{50}}{d_{50}} = 3 - 4.5 \tag{23}$$

http://dx.doi.org/10.5772/intechopen.69167

Sherard et al. [7]:

$$\frac{D_{15}}{d_{85}} < 9 \tag{24}$$

where D and d denote the filter and the base soil, respectively.

Rules for broadly graded soils

Terzaghi's [10, 11]:

$$\frac{D_{15}}{d_{85}} \leq 4, \frac{D_{15}}{d_{15}} \geq 4 \tag{25}$$

US Bureau of Reclamation [12]:

$$\frac{D_{50}}{d_{50}} = 12 - 58, \frac{D_{15}}{d_{15}} = 12 - 4 \tag{26}$$

Bertram [13]:

$$\frac{D_{15}}{d_{85}} \leq 5, \frac{D_{15}}{d_{15}} = 5 - 9 \tag{27}$$

Cistin [14]:

$$\frac{D_{10}}{d_{60}} < 5, U_D = \frac{D_{60}}{D_{10}} < 5 \tag{28}$$

where D and d are diameters of the filter soil and the base soil.

Acknowledgements

The chapter is related to the work of the Research Project OTKA 1457/86 on river dykes and of the Research Project NKFP B1 2006 08 on MSW landfills.

Author details

János Lőrincz[1], Emöke Maria Imre[2]* and Vijay Pal Singh[3]

*Address all correspondence to: imreemok@gmail.com

1 Geotechnical Department, BME, Budapest, Hungary

2 Power System Department, KVK Faculty, Obuda University of Budapest

3 Department of Biological & Agricultural Engineering, Zachry Department of Civil Engineering, Water Management & Hydrological Science, Texas A&M University, USA

References

[1] Bonelli S, Nicot F, editors. Erosion in Geomechanics Applied to Dams and Levees. London Wiley-ISTE; 2013

[2] Van Beek VM, Knoeff H, Sellmeijer H. Observations on the process of backward erosion piping in small-, medium- and full-scale experiments. European Journal of Environmental and Civil Engineering. 2011;**15**:1115-1137

[3] Sellmeijer H, de la Cruz JL, van Beek VM, Knoeff H. Fine-tuning of the backward erosion piping model through small-scale, medium-scale and IJkdijk experiments. European Journal of Environmental and Civil Engineering. 2011;**15**:1139-1154

[4] Khomenko, V.P. Suffosion hazard: Today's and tomorrow's problem for cities. In Proceedings of IAEG2006, Nottingham, UK, 6-10 September 2006; No. 577, p. 8-18

[5] Lubockov EA. The calculation of suffosion properties of non-cohesive soils when using the non-suffosion analogue. In: Proceedings of the International Conference on Hydraulic Research; Brno; 1965. pp. 135-148

[6] Kenney TC, Lau D. Internal stability of granular filters. Canadian Geotechnical Journal. 1985;**22**:215-225

[7] Sherard JL, Dunningan LP, Talbot JR. Basic properties of sand and gravel filters. Journal of Geotechnical Engineering. 1984;**110**:684-700

[8] Cedergren HR. Seepage control in earth dams. In: Embankment-Dam Engineering. New York: Wiley; 1973. pp. 21-45

[9] Schuler U, Brauns J. Behaviour of coarse and well-graded filters. Filters in Geotechnical and Hydraulic Engineering: Proceedings of the 1st international conference 'Geo-filter', Karlsruhe, Germany, 20-22 October. 1993 by J. Brauns (Editor), M. Heibaum (Editor), U. Schuler (Editor), CRC Press. Karlsruhe, Germany, pp. 3-18

[10] Fannin J, Terzaghi K. From theory to practice in geotechnical filter design. Journal of Geotechnical and Geoenvironmental Engineering. 2008;**134**:267-276

[11] Terzaghi K, Peck RB, Mesri G. Soil Mechanics in Engineering Practice. New York: Wiley; 1996

[12] US Bureau of Reclamation. Earth Manual Part 1. 3rd ed. Denver: U.S. Department of the Interior Bureau of Reclamation. Geotechnical Research Technical Service Center; 1998

[13] Bertram GE. An Experimental Investigation of Protective Filters. Harvard Soil Mechanics Series No. 7. Cambridge: Graduate School of Engineering, Harvard University; 1940

[14] Cistin, J. Zum Problem mechanischer Deformationen nichtbindiger Lockergesteine durch die Sickerwasserströmung in Erddämmen. Wasserwirtschaft Wassertechnik. 1967;**2**:45-49 (In German)

[15] Lőrincz J. Grading entropy of soils. Budapest: University Doctoral Thesis, Technical University of Budapest; 1986 (In Hungarian)

[16] Singh VP. Entropy Theory in Hydraulic Engineering: An Introduction. ASCE Press; 2014. ISBN 978-0-7844-1272-5

[17] Lőrincz J. On particle migration with the help of grading entropy. Filters in Geotechnical and Hydraulic Engineering: Proceedings of the 1st international conference 'Geo-filter', Karlsruhe, Germany, 20-22 October., 1993 by J. Brauns (Editor), M. Heibaum (Editor), U. Schuler (Editor). CRC Press. Karlsruhe, Germany. pp. 63-66

[18] Lőrincz J. On granular filters with the help of grading entropy Filters in Geotechnical and Hydraulic Engineering: Proceedings of the 1st international conference 'Geo-filter', Karlsruhe, Germany, 20-22 October. Karlsruhe, Germany, 1993 by J. Brauns (Editor), M. Heibaum (Editor), U. Schuler (Editor). CRC Press. pp. 67-70

[19] Imre E. Characterization of dispersive and piping soils. In: Proceedings of XI. European Conference on Soil Mechanics and Foundation Engineering, Danish Geotechnical Society Copenhagen. Vol. 2; Copenhagen. 1995. pp. 49-55

[20] Imre E, Lőrincz J, Szendefy J, Trang PQ, Nagy L, Singh VP, Fityus S. Case studies and benchmark examples for the use of grading entropy in geotechnics. Entropy. 2012;**14**:1079-1102

[21] Lőrincz J, Imre E, Fityus S, Trang PQ, Tarnai T, Talata I, Singh VP. The grading entropy-based criteria for structural stability of granular materials and filters. Entropy. 2015;**17**: 2781-2811

[22] Imre E, Nagy L, Lőrincz J, Rahemi N, Schanz T, Singh VP, Fityus S. Some comments on the entropy-based criteria for piping. Entropy. 2015;**17**:2281-2303

[23] Korn GA, Korn TM. Mathematical Handbook for Scientists and Engineers. 2nd ed. New York: McGraw-Hill; 1975

[24] Einav I. Breakage mechanics—Part I: Theory. Journal of the Mechanics and Physics of Solids. 2007;**55**(3):1274-1297

[25] Kézdi Á. Phase Movements in Granular Soils. Notes of Budapest University of Technology and Economics, Graduate Courses. Budapest: Budapest University of Technology and Economics; 1975 (manuscript)

[26] O'Sullivan C. Applying micro-mechanical analysis to realistic sands. In: Proceedings of the International Symposium on Discrete Element Modelling of Particulate Media: In Celebration of the 70 Birthday of Colin Thornton, Birmingham, 29-30 March 2012. Royal Society of Chemistry Publishing. London.

[27] Zhang LM, Chen Q. Seepage failure mechanism of the Gouhou rock-fill dam during reservoir water infiltration. Soils Found. 2006;**46**:557-568

[28] Gouhou Dam Failure Investigation Team. Technical details of the Gouhou Dam. In: Gouhou Concrete-Faced Rockfill Dam—Design, Construction, Operation, and Failure. Beijing: Water Conservancy and Hydropower Press; 1996. pp. 111-245

[29] Lőrincz J, Imre E, Trang PQ, Telekes G, Juhász M, Fityus S. Grading entropy and degradation of sands. In: XV Danube - European Conference on Geotechnical Engineering. Vienna. 9-11. September 2014. *Austrian Society of Engineers and Architects*, ÖIAV. pp. 717-722

[30] Emerson WW. A classification of soil aggregates based on their coherence in water. Australian Journal of Soil Research. 1967;**5**:47-57

[31] Rengasamy P, Greene RSB, Ford GW. The role of clay fraction in the particle arrangement and stability of soil aggregates—A review. Clay Research. 1984;**3**:53-67

[32] Yong RN, Amar IS, Harald PL, Jorgensen MA. Interparticle action and rheology of dispersive clays. Journal of Geotechnical Engineering. 1979;**105**:1193-1211

[33] ASTM 2005 C136-06. Standard Test Method for Sieve Analysis of Fine and Coarse Aggregates. West Conshohocken: American Society for Testing and Materials; 2003